Finite Elements
in Fluids —Volume 1

Finite Elements in Fluids—Volume 1

Viscous Flow and Hydrodynamics

Edited by

R. H. Gallagher
Civil Engineering, Cornell University, Ithaca, N.Y.

J. T. Oden
Engineering Mechanics, University of Texas

C. Taylor
Civil Engineering, University College Swansea

and

O. C. Zienkiewicz
Civil Engineering, University College Swansea

A Wiley–Interscience Publication

JOHN WILEY & SONS
London · New York · Sydney · Toronto

Library of Congress Cataloging in Publication Data:

International Conference on the Finite Element Method in
Flow Analysis, University College of Wales, 1974. Finite
elements in fluids.

'A Wiley–Interscience publication.'
CONTENTS: v. 1. Viscous flow and hydrodynamics.—
v. 2. Mathematical foundations, aerodynamics, and
lubrication.
1. Fluid dynamics—Congresses. 2. Finite element method—
Congresses. I. Gallagher, Richard H., ed. II. Title.

QA911.I53 1974   532'.051   74-13573

ISBN 0 471 29045 9

Printed in Great Britain by J. W. Arrowsmith Ltd.,
Bristol BS3 2NT.

Contributing Authors

J. C. M. BERKHOFF   *Delft Hydraulics Laboratory,*
Mathematical Branch, Laboratory De Voorst,
Emmeloord, North-East Polder, Netherlands.

J. M. DAVIS   *Department of Civil Engineering,*
University College of Wales,
Singleton Park, Swansea, SA2 8PP, U.K.

DR. C. S. DESAI   *Department of Civil Engineering,*
Virginia Polytechnic Institute and State University,
Blacksburg, Virginia, 24061, U.S.A.

PROFESSOR   *Department of Structural Engineering,*
R. H. GALLAGHER   *Cornell University, Ithaca, New York, 14850,*
U.S.A.

G. HIRIART   *Department of Mechanical Engineering,*
U.S. Navy Postgraduate School, Monterey,
California 93940, U.S.A.

K. H. ICEMAN   *Staff Engineer, Water Resources Engineers,*
710 S. Broadway, Walnut Creek, California,
94596, U.S.A.

K. ITO   *Institute of Industrial Science,*
University of Tokyo, Roppongi, Minato-Ku,
Tokyo 106, Japan.

DR. I. P. KING   *Resource Management Associates,*
3706 Mt. Diablo Blvd., Suite 200,
Lafayette, CA 94549, U.S.A.

DR. R. W. LEWIS   *Department of Civil Engineering,*
University College of Wales, Singleton Park,
Swansea, SA2 8PP, U.K.

DR. S. P. NEUMAN   *Senior Scientist, Institute of Soils and Water*
Agricultural Research Organization,
P.O. Box 6, Bet Dagan, Israel.

PROFESSOR   *Department of Mechanical Engineering,*
R. E. NEWTON   *U.S. Navy Postgraduate School, Monterey,*
California, 93940, U.S.A.

W. R. NORTON   *Resource Management Associates,*
3706 Mt. Diablo Blvd., Suite 200,
Lafayette, CA 94549, U.S.A.

T. OHTSUBO — *Japan Advanced Numerical Analysis, Inc., Kamiosaki, Shinagawa-ku, Tokyo 141, Japan.*

PROFESSOR M. D. OLSON — *Department of Civil Engineering, The University of British Columbia, Vancouver 8, B.C., Canada.*

PROFESSOR T. SARPKAYA — *Department of Mechanical Engineering, U.S. Navy Postgraduate School, Monterey, California, 93940, U.S.A.*

T. TAMANO — *Nippon Steel Corporation, Tokyo, Japan.*

DR. C. TAYLOR — *Department of Civil Engineering, University College of Wales, Singleton Park, Swansea, SA2 8PP, U.K.*

DR. E. VERNER — *Department of Civil Engineering, University of Texas at Austin, Austin, Texas 78731, U.S.A.*

DR. W. VISSER — *Koninklijke/Shell Exploration en Produktie Laboratorium, Rijswijk, The Netherlands.*

M. VAN DER WILT — *National Lucht-en Ruimtevaart Laboratorium, Amsterdam, The Netherlands.*

PROFESSOR Y. YAMADA — *Institute of Industrial Science, University of Tokyo, Roppongi, Minato-Ku, Tokyo 106, Japan.*

DR. Y. YOKOUCHI — *Institute of Industrial Science, University of Tokyo, Roppongi, Minato-Ku, Tokyo 106, Japan.*

PROFESSOR O. C. ZIENKIEWICZ — *Department of Civil Engineering, University College of Wales, Singleton Park, Swansea, SA2 8PP, U.K.*

Preface

The finite element method, which was first devised as a procedure for structural analysis, has come to be recognized as an effective analysis tool for a wide range of physical problems. Among these are problems in the field of flow analysis, a subject which is herein interpreted to encompass not only the flow of fluids but also heat flow and lubrication phenomena.

The application of the finite element method to flow analysis problems is a relatively recent development but, nevertheless, a significant literature on the topic has already emerged. Also, a formidable amount of research and application work, involving many aspects of the subject, was in progress by early 1972, the point in time when it was decided to initiate arrangements for a Symposium on finite element methods in flow problems. In planning the Symposium the decision was therefore made to provide a forum for three different types of paper. One type was to be the invited state-of-the-art review of a major aspect of finite element flow analysis. Another type was to feature the in-depth treatment of a particular problem. Finally, there was the paper that outlined work completed, or in progress, on a particular problem.

The International Symposium on the Finite Element Method in Flow Problems was organized along the lines described above and was held at University College of Wales, Swansea, Jan 7–11, 1974. Over 70 papers were presented in the three categories and, despite difficulties in transportation and labour relations that prevailed at the time, a capacity attendance of 250 participants were recorded. A proceedings, available at the time of the Conference, was published by the University of Alabama Press at Huntsville under the title, *Finite Element Methods in Flow Problems*. This contained full-length papers in the third category described above and extended abstracts of papers in the first two categories.

Papers in the first two categories are found in this book and in the companion, *Finite Elements in Fluids. Volume 2, Mathematical Foundations, Aerodynamics and Lubrication*.

Although the division of the contents of these two books is accurately characterized by the topic designations (viscous flow, hydrodynamics, mathematical foundations, aerodynamics, lubrication) it would also be possible to assign the designations 'engineering applications' and 'theoretical basis' to them. Exceptions to both modes of designation can be readily identified by the reader.

It is natural to inquire, with respect to a field in which numerical methods in the form of finite difference and series solution techniques have a well-established place, 'Why Finite Elements?'. This question is posed and answered by Zienkiewicz in the opening chapter of this book. Using it as a vehicle for a fundamental review of the various perspectives of the finite element method, he branches out to a summary of the key theoretical aspects of the method and then to a description of many avenues of future investigation in finite element analysis. Readers with a bias towards either the practical content of the present volume or the more theoretical content of the companion volume should find this a useful introduction to the subject.

The thirteen chapters which follow the opening chapter are broadly categorized under the titles of 'viscous flow', encompassing Chapters 2–4, and 'hydrodynamics', Chapters 5–14; but still further subdivision is called for in the latter. These subdivisions might loosely be referred to as 'large body of water analysis', (Chapters 5–7, 'porous flow' (Chapters 8–10) and 'wave problems' (11–14).

In the work on viscous, incompressible flow, Zienkiewicz and Godbole first present in Chapter 2 a general framework for the finite element analysis of this type of problem and illustrate its application to both fluid and material flow situations. Olson (Chapter 3) combines the relevant equations in such a way as to produce a single fourth order differential equation in the stream function whose solution by the finite element method has the features of plate bending analysis in structural mechanics. Yamada and his colleagues return in Chapter 4 to the approach taken by Zienkiewicz and Godbole of formulating the problem directly in terms of the basic, lower order equations. They deal with the analysis of material flow, using a rather different approach, and investigate various approaches.

Because of its role in the most critical pollution phenomena, the analysis of dispersion in estuaries holds a special place in current numerical flow analysis work. Taylor and Davis open the section that concerns large bodies of water with a finite element treatment of this problem and of the allied problem of tidal propagation in estuaries. Gallagher (Chapter 6) treats lake circulation under the action of surface shear due to wind in a manner that accounts for the variable depth of the lake and discusses simplified thermal analysis models which seek to predict the annual cycle of stratified temperature change of the lake. The question of stratification, which couples density and temperature variations across the depth of large bodies of water, is examined in Chapter 7 by King, Norton and Iceman. They adopt an approach quite similar to that taken in Chapters 2 and 4, except that now a convective diffusion equation with density as the independent variable is coupled to the equations of viscous, incompressible flow.

To open the part of the book concerned with flow through porous media, Desai (Chapter 8) gives a well-referenced review of the state-of-the-art. Then, in Chapter 9, Lewis, Verner and Zienkiewicz examine the problem of two-phase flow in porous media. A prominent application of this is the displacement of oil from the pores of a hydrocarbon-bearing medium due to the injection of an immiscible fluid. Solutions to this problem by means of the finite element approach are described. Neuman (Chapter 10), in addressing the topic of seepage in unsaturated and partly-saturated porous media emphasizes applications to dams, but also treats discharge to a well.

The two opening papers of the part of the book containing problems loosely classified as 'wave problems' deal with floating bodies. Newton (Chapter 11) examines oscillatory motions of such bodies and establishes practical guidelines for dealing with the fluid region which is actually of infinite dimension. Visser and van der Wilt, in Chapter 12, perform transient analyses of floating structures with a mind to calculation of the forces acting on offshore structures. Berkhoff (Chapter 13) deals principally with the finite element analysis of the action of water waves on stationary objects and in this work first gives separate attention to the diffraction and refraction problems and then to their combination. Finally, in a paper which does not in fact deal with waves but rather with a problem of profound practical design importance for fluid machinery, Sarpkaya and Hiriart (Chapter 14) describe solutions for the deflection of a free jet by a solid boundary.

It can be said that, as a group, the papers in this book emphasize the construction of finite element representations for practical problems for which the governing equations are well known and widely used. The construction is most often based on the Galerkin approach, but variational concepts find a place. The actual geometric forms of the elements and the descriptions of the variables within the elements follow the established and familiar lines that these have taken in structural mechanics. In contrast to structural mechanics, however, nearly all practical problems in fluid flow analysis include non-linear terms and for this reason the solution of non-linear algebraic equations occupies a prominent place in many of the papers included herein. Also, the scope of practical problems which are described and their comparison with available field data and alternative solutions is relatively wide.

A certain commonality prevails in the outlook of the authors of papers in this volume and for that reason it has been possible to achieve a unity of notation to a limited extent. Such coordination is infeasible in the companion volume due to the disparate nature of the contributions found therein.

Acknowledgements
The editors wish to express their appreciation of the financial support of A.F.O.S.R. and the cooperation of the International Journal of Numerical Methods in Engineering, the Institute of Mathematics and its Applications,

the International Journal of Computers and Fluids, and the International Association for Hydraulic Research. The above support and cooperation helped in making the Symposium a success and in achieving its objectives.

List of Symbols

Although it is not possible to effect complete unity of nomenclature in a work involving fourteen different groups of authors an attempt has been made to coordinate the meaning of certain symbols. These are given below. Other symbols are defined where they appear in the text. Matrices and vectors are symbolized by means of bold face type.

A	Cross-sectional area
$\mathbf{A}$	Acceleration of a point in a fluid
$\mathbf{a}, a_i$	Unknown-parameter vector and component
a	Half-amplitude of wave
$\mathbf{b}$	Undetermined parameters
C^{m-1}	Designation of continuity condition of order $m-1$
C_c	Chezy friction coefficient
C_f	Skin friction coefficient
C_T	Thrust coefficient (Chapter 14)
c^{-i}	Concentration factor
$\mathscr{D}(\ )$	Differential operator
$\mathbf{D}$	Matrix of elastic constants
D	Dispersion coefficient (Chapter 5)
D_x, D_y	Turbulent diffusion coefficient
e	Eddy diffusion coefficient
$F(\boldsymbol{\sigma})$	Yield condition
$\mathbf{F}, F_j$	Force vector and component
$\mathscr{G}(\ )$	Constraint condition
g	Acceleration due to gravity
H	Depth from water level to bed
$\bar{H}$	Amplitude of wave $(= 2a)$
h	Depth from mean level of water to surface
k	Permeability
L_1, L_2, L_3	Area coordinates
l_x, l_y	Direction cosines
$\mathbf{N}, N_j$	Vector of shape functions and component
p	Pressure
Q	Flow input
q	Source term
Re	Reynolds number

r	Radial coordinate
$\mathbf{S}$	Deviatoric stress
s	Dissipation factor per unit volume
T	Temperature
$\overline{\mathbf{T}}$	Vector of prescribed surface tractions
t	Time
$U_0 V$	Averaged velocities
W_x, W_x	Components of wind force on water surface
W_i	Weighting function
u, v, w	Cartesian components of velocity
v_g	Group velocity of waves
$\mathbf{X}, X, Y, Z$	Body force vector and Cartesian components
α	Penalty factor
Γ	Surface variable
γ	Unit weight
ε	Strain (Chapter 2); turbulent exchange coefficient (Chapter 7)
η_R	Reverse-thrust ratio (Chapter 14)
θ	Moisture content expressed as volume fraction
Λ	Particular solution (Chapter 2); Coriolis parameter (Chapters 5, 6)
λ	Lagrange multiplier
μ	Viscosity
ν	Kinematic viscosity
ξ	Vorticity
Π	Functional
ρ	Density
$\boldsymbol{\sigma}$	Stress vector
τ	Shear stress
υ	Porosity
ϕ	Variable (hydraulic head, velocity potential etc)
ψ	Stream function
Ω	Domain of problem (volume, area etc.)
ω	Frequency

Contents

Chapter 1

Why Finite Elements?

O. C. Zienkiewicz

1.1 Introduction

What are finite elements? Why finite elements when to date we have been served adequately by finite difference methods? These may well be the questions asked by a fluid mechanician who, in the search for a numerical solution of his complex problems, governed by well defined differential equations, has developed powerful finite difference codes which today are widely used in his practice!

To supply an answer to this, certainly non-trivial, question we shall try, in this introduction, to

(a) define the finite element method in its general form, and
(b) show its different facets which to a greater or lesser extent have been used to date.

We shall see how, within a broad definition, the finite difference techniques fall into a 'subclass' of the general finite element methodology which indeed embraces many other classical approximation procedures. This generalization of the finite element concept is by no means a 'power bid' by its over-enthusiastic adherents. On the contrary, it serves, we believe, to lay a firm foundation to a wide variety of solution methods and provide expanded possibilities of application.

Before embarking on the main theme of this introduction, the reader may well ask the further question: 'Is the effort of reading this exposé, and indeed the proceedings of this conference, and acquainting himself with a new field, worthwhile?' Will it in fact provide tools which are in all respects superior to those to which he is accustomed? The answer must be left to his intuition. However, the observation of the field of structural and solid mechanics, in which finite element procedures have today 'taken over' other alternatives, may provide a clue. The research developments of the last fifteen years (since the first mention of the 'finite element' name was made[1,2]) have become everyday practice in many stress analysis/structures situations. It is, by analogy, probable that in the field of fluid mechanics a similar revolution is now feasible.

1.2 The finite element concept

The finite element method is concerned with the solution of mathematical or physical problems which are usually defined in a continuous domain either by local differential equations or by equivalent global statements. To render the problem amenable to numerical treatment, the infinite degrees of freedom of the system are *discretized* or replaced by a finite number of unknown parameters, as indeed is the practice in other processes of approximation.

The original 'finite element' concept replaces the continuum by a number of subdomains (or elements) whose behaviour is modelled adequately by a limited number of degrees of freedom and which are *assembled* by processes well known in the analysis of discrete systems. Often at this early stage the model of the element behaviour was derived by a simple physical reasoning avoiding the mathematical statement of the problem. While one can well argue that such an approach is just as realistic as formal differential statements (which imply the possibility of an infinite subdivision of matter), we prefer to give here a more general definition embracing a wider scope.*

Thus we define the finite element process as any approximation process in which

(a) the behaviour of the whole system is *approximated* to by a finite number, n, of parameters, a_i, $i = 1 - n$ for which
(b) the n-equations governing the behaviour of the whole system, i.e.

$$F_j(a_i) = 0 \qquad j = 1 - n \qquad (1.1)$$

can be assembled by the simple process of addition of terms contributed from all subdomains (or elements) which divide the system into physically identifiable entities (without overlap or exclusion). Thus

$$F_j = \sum F_j^e \qquad (1.2)$$

where F_j^e is the element contribution to the quantity under consideration.

This broad definition allows us to include in the process both physical and mathematical approximations and, if the 'elements' of the system are simple and repeatable, to derive prescriptions for calculation of their contributions to the system equations which are generally valid. Further, as the process is precisely analogous to that used in discrete system assembly, computer programs and experience accumulated in dealing with discrete systems can be immediately transferred.

An important practical point of the approximation has been specifically excluded here. This concerns the fact that often contributions of the elements

---

* In some situations, such as for instance the behaviour of granular media, the first approach is still one of the most promising as continuously defined constitutive relations have not yet been adequately stated.

are highly localized and only a few non-zero terms are contributed by each element. In practice this localization results in sparse and often banded equation systems, reducing computer storage requirements. Whilst most desirable in practice this feature is not essential to the definition of the finite element process.

What then are the procedures by which a finite element approximation can be made? We have already mentioned—but now exclude from further discussion here—the *direct physical approach* and will concentrate on any problem which can be defined mathematically *either* by a (set) of differential equations valid in a domain Ω

$$\mathscr{D}(\phi) = 0 \tag{1.3}$$

together with their associated boundary conditions on boundaries (Γ) of the domain

$$\mathbf{B}(\phi) = 0 \tag{1.4}$$

or by a variation principle requiring stationarity (max., min. or 'saddle') of some scalar functional Π

$$\Pi = \int_\Omega G(\phi)\, d\Omega + \int_\Gamma g(\phi)\, d\Gamma \tag{1.5}$$

In both statements ϕ represents either the single unknown function or a set of unknown functions.

To clarify ideas, consider a particular problem presented by seepage flow in a porous medium where ϕ is the hydraulic head (a scalar quantity).

The specific governing equation is now written for a two-dimensional domain Ω as:

$$\mathscr{D}(\phi) \equiv \frac{\partial}{\partial x}\left(k\frac{\partial \phi}{\partial x}\right) + \frac{\partial}{\partial y}\left(k\frac{\partial \phi}{\partial y}\right) + Q = 0 \tag{1.6}$$

together with boundary conditions

$$B(\phi) \equiv \phi - \bar{\phi} \qquad \text{on } \Gamma_1$$

$$B(\phi) \equiv k\frac{\partial \phi}{\partial n} - q \qquad \text{on } \Gamma_2 \tag{1.7}$$

in which both k (the permeability) and Q (the flow input) may be functions of position (and, in non-linear problems, of the gradients or values of ϕ). (In the above, all the quantities are scalars and the bold notation for vectors is not used.)

An alternative formulation (for linear problems) requires stationarity (a minimum) of a functional

$$\Pi = \int_{\Omega} \left\{ \tfrac{1}{2}k\left(\frac{\partial \phi}{\partial x}\right)^2 + \tfrac{1}{2}k\left(\frac{\partial \phi}{\partial y}\right)^2 - Q\phi \right\} d\Omega - \int_{\Gamma_2} q\phi \, d\Gamma \qquad (1.8)$$

for ϕ which satisfies only the first boundary condition.

In general if a functional Π exists then an associated set of (Euler) differential equations can always be found but the reverse is not necessarily true.

To obtain a finite element approximation to the general problem defined by Equations 1.3 to 1.5 we proceed as follows:

(a) the unknown function is expanded in a finite set of assumed, known trial functions N_i and unknown parameters a_i, i.e.

$$\hat{\phi} = \sum N_i a_i = \mathbf{N}\mathbf{a} \qquad (1.9)$$

and

(b) the approximation must be cast in a form of n equations *which are defined as integrals over* Ω *and* Γ i.e.

$$\mathbf{F}_j = \int_{\Omega} \mathbf{E}(\hat{\phi}) \, d\Omega + \int_{\Gamma} \mathbf{e}(\hat{\phi}) \, d\Gamma \qquad (1.10)$$

$j = 1 - n$.

Immediately we note that the basic definitions of the finite element process previously given apply, as *for integrable functions*

$$\int_{\Omega} (\quad) \, d\Omega \equiv \sum \int_{\Omega^e} (\quad) \, d\Omega \qquad (1.11)$$

and

$$\int_{\Gamma} (\quad) \, d\Gamma \equiv \sum \int_{\Gamma^e} (\quad) \, d\Gamma \qquad (1.12)$$

in which Ω^e, Γ^e represent 'element' subdomains.

The problem of how the integrals of approximations are formed is thus the first, crucial, question of casting a problem in a finite element form.

1.3 Approximation integrals

1.3.1 Variational principles

If the problem is stated in terms of a stationary functional Π then the formulation is most direct. We can write the approximate form of the functional as

$$\Pi = \hat{\Pi} = \Pi(\hat{\phi}) \qquad (1.13)$$

and for stationarity we have a set of equations

$$\mathbf{F}_j \equiv \frac{\partial \Pi}{\partial \mathbf{a}_j} = 0 \qquad (1.14)$$

which by definition of Π is already cast in an integral form. This basis of forming a finite element approximation has been and remains most popular, *providing* a physically meaningful variational principle exists and can be readily identified. This has led to statements of the kind that the finite element method is a 'variational process' which is, however, too limited a definition, as other alternatives, often more powerful, are present. The important question of how to proceed from the differential equation directly in cases where a variational principle does not exist or cannot be identified remains. The answer to it lies in the reformulation by use of weighting function, or by the introduction of 'pseudo-variational' principles.

1.3.2 Weighted integral statements

It is obviously possible to replace the governing equations (1.6 or 1.7) by an integral statement in all respects equivalent, i.e.

$$\int_\Omega \mathbf{W}^T \mathscr{D}(\phi) \, d\Omega + \int_\Gamma \overline{\mathbf{W}}^T \mathbf{B}(\phi) \, d\Gamma \qquad (1.15)$$

in which $\mathbf{W}$ and $\overline{\mathbf{W}}$ are completely arbitrary, 'weighting' functions. Immediately an approximation is possible in an integral form by choosing specific functions W_j and $\overline{W}_j$ and writing[3,4]

$$\mathbf{F}_j = \int_\Omega \mathbf{W}_j^T \mathscr{D}(\hat{\phi}) \, d\Omega + \int_\Gamma \overline{\mathbf{W}}_j^T \mathbf{B}(\hat{\phi}) \, d\Gamma \qquad (1.16)$$

The process is known as the weighted residual method if $\mathscr{D}(\hat{\phi})$ and $\mathbf{B}(\hat{\phi})$ are recognized as *residuals* by which the approximation misses the zero value required. Classical procedures of Galerkin's method, collocation etc. are immediately recognized. The Galerkin process, in which the weighting function and the trial function are identical ($W_j \equiv N_j$), is the most popular form used in finite element analysis. In many chapters of this book we shall discover its application.

Either form of deriving integral statements and hence the set of approximating equations can be and has been used in practice. The variational principle possesses however a unique advantage. If the function is quadratic in $\mathbf{a}$ the set of approximating equations (1.14) can be written as

$$\mathbf{Ka} + \mathbf{P} = 0 \qquad (1.17)$$

in which $\mathbf{K}$ is always a symmetric matrix ($K_{ij}^T = K_{ji}$). For linear differential equations the weighting processes will also result in a similar set of equations via Equation 1.16; however, these will not in general be symmetric. The user of finite difference procedures may well be acquainted with such dissymmetries which often present computational difficulties. This symmetry can indeed be shown to be a precondition for the existence of a variational principle—it will indeed be found that the Galerkin's method of weighting will yield identical equations as those derived from a variational principle whenever this exists.

Because of this (and certain other) advantages of variational formulations, much work of a theoretical nature has been put in to establish equivalent functionals for problems defined by differential equations or to create pseudo-variational functionals.[5,6,7,8] (See Chapters 3 and 12 for such functionals.)

1.3.3 Pseudo-variational principles. Constraints by Lagrange multipliers or penalty functions. Adjoint variables and least square processes

Pseudo-variational principles can be established by various means. These include constrained variational principles and the extreme cases obtained from these by the use of adjoint functions, or the application of least square methods.

Constrained variational principles require that some functional Π be stationary and subject to constraints say of the type given by some differential relations

$$\mathscr{G}(\phi) = 0 \quad \text{in } \Omega \tag{1.18}$$

In such cases we can proceed to establish a new variational principle in either of two ways. In the first we introduce an additional set of functions λ known as *Lagrangian* multipliers and require

$$\bar{\Pi} = \bar{\Pi}\left(\left\{\begin{matrix} \phi \\ \lambda \end{matrix}\right\} = \Pi + \int_\Omega \lambda^T \mathscr{G} \, d\Omega\right) \tag{1.19}$$

to be stationary. The variation of this functional results in

$$\delta\bar{\Pi} = \delta\Pi + \int_\Omega \delta\lambda^T \mathscr{G} \, d\Omega + \int_\Omega \lambda^T \delta\mathscr{G} \, d\Omega \tag{1.20}$$

which can only be true if both the stationarity of Π is stationary and the constraints (1.18) are satisfied.

The use of Lagrangian multipliers in practice is somewhat limited, owing to two drawbacks. First, the additional functions λ have to be discretized thus requiring a larger number of unknowns in the final problem. Second, it will always be found that, if Π is quadratic and $\mathscr{G}$ a linear equation, the final discretized form Equation 1.17 has zero diagonal terms corresponding to the parameters discretizing λ (this is obvious from the inspection of Equation 1.20).

To obviate some of the difficulties associated with the use of Lagrangian multipliers it is possible to require the stationarity of a modified functional based on a *penalty function*. For, if at the solution we require a simultaneous satisfaction of the stationarity of Π *and* the satisfaction of constraints we can minimize approximately

$$\bar{\Pi} = \Pi + \alpha \int_{\Omega} \mathscr{G}^T \mathscr{G} \, d\Omega \qquad (1.21)$$

in which α is some large (positive) number 'penalizing' the error of *not* satisfying the constraints. As no procedure is without a drawback, we note here a purely numerical difficulty: as α becomes large the discretized equations tend to become ill-conditioned. However, with modern computers and high precision arithmetic, penalty function operations are becoming increasingly popular and their use more widespread.[9] In Chapter 2 we show how such procedures can be used effectively in fluid mechanics.

What if even a constrained variational principle does not appear to be identified? Clearly *both* methods given above are still applicable by putting $\Pi \equiv 0$ and identifying the constraints with the full set of differential equations to be satisfied. Thus we can either make

$$\bar{\Pi} = \bar{\Pi} \binom{\phi}{\lambda} = \int_{\Omega} \lambda^T \mathscr{G} \, d\Omega \qquad (1.22)$$

stationary or alternatively minimize

$$\bar{\Pi} = \int_{\Omega} \mathscr{G}^T \mathscr{G} \, d\Omega \qquad (1.23)$$

The first is equivalent to the use of *adjoint functions*[10] while the latter is the straightforward application of the *least squares* procedures of approximation.[3,4]

The pseudo-variational principle established by Equation 1.22 in which a new, adjoint, function λ is introduced is of little practical use. The resulting discretized equation systems for parameters defining approximations to ϕ and λ are entirely uncoupled and indeed there is little virtue in the symmetry arising from the whole system as a zero diagonal exists throughout. Nevertheless this approach gives another interpretation of the Galerkin weighting process if similar expansions are used for ϕ and λ. The least square formulation, on the other hand, results in well conditioned equation systems and deserves much wider attention in the finite element literature than it has so far received.[11,12]

1.3.4 Direct integral statements—virtual work

In many physical situations it is possible to formulate the problem directly in an integral form avoiding the necessity of writing down the full governing differential equations. In particular the *principle of virtual work* in mechanics can be so stated with greater generality than that arising from differential equations. Indeed in such cases the weighted residual form given by Equation 1.15 arises in a form which can be obtained from such equations by the use of *integration by parts*. Such integration reduces the continuity requirement imposed on both functions W and N by 'integrability' (to which we shall refer in the next section). This relaxation of requirements is known mathematically as a 'weak formulation' of the problem. It is of philosophical interest to interject here a thought that perhaps such weak formulations are indeed the requirements of Nature as opposed to differential equations which, at certain physical discontinuities, are meaningless.

In structural mechanics, virtual work principles have almost replaced the formulation based directly on energy statements because of the wider applicability of virtual work (and because they avoid often complex algebraic manipulations). In Chapter 2 the author shows how such statements form an extremely realistic and simple approach to fluids.[13]

Table 1.1   Finite element approximation

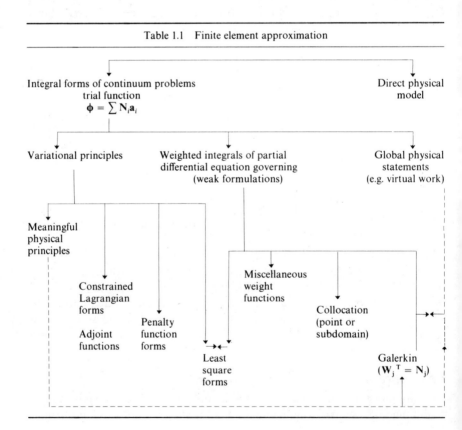

Table 1.1 summarizes the basic processes by which the integral forms of approximation can be made as a preliminary to finite element analysis.

1.4 Partial discretization

At this point it is appropriate to mention that it is often convenient to discretize the problem only partially in a manner which, say, reduces a set of differential equations in three independent variables not directly to a numerical set of equations but to a lower order differential equation, say, with only one variable. This first differential equation can then, on occasion, be solved more efficiently by exact procedures or alternative numerical solutions.

Such 'partial discretization' is particularly useful if the 'shape' of the domain in one of the independent directions is simple. This may arise if prismatic or axisymmetric shapes are considered in a three-dimensional problem or if one of the dimensions is that of *time*.

Considering the last case as a concrete example the trial function expression discretizing the unknown ϕ

$$\phi = \phi(x, y, z, t) \tag{1.24}$$

is made by modifying Equation 1.9 to

$$\phi = \sum N_i a_i = \mathbf{N}\mathbf{a} \tag{1.25}$$

in which

$$N_i = N_i(x, y, z) \tag{1.26}$$

i.e. is only a function of position and $\mathbf{a}$ is now a set of parameters which are a function of time

$$\mathbf{a} = \mathbf{a}(t) \tag{1.27}$$

'Partial variations' of variation principles (Equation 1.13) or the use of any of the weighting procedures (Equation 1.16) in which the weighting functions do not include the independent variable t can now be made, reducing the formulation to a *set of ordinary differential equations*.

In fluid mechanics and flow problems we shall often find such a discretization useful and the ordinary set of differential equations can often be solved efficiently by simple finite difference schemes as well as by a secondary application of the finite element methodology.[1]

1.5 Trial functions

1.5.1 General principles

So far, beyond mentioning that the unknown function ϕ is expanded as in Equation 1.9 by a set of trial functions $\mathbf{N}$, no specific mention was made of the form these trial functions should take or what limitations have to be imposed on them. We shall here consider, in very general terms, some of the guidelines, though by necessity the discussion cannot do full justice to this problem which is crucial to the success of the finite element process. For details, therefore, the reader is referred to the text[1] and to numerous other publications in which differential trial (or shape) functions are discussed.

As the trial functions **N** are constructed for practical reasons in a *piecewise manner*, i.e. using a different definition within each 'element', the question of required interelement continuity is important in their choice. This continuity has to be such that *either* the integrals of the approximation given in general by Equation 1.10 or in particular forms by Equations 1.14 and 1.16 can be evaluated directly, without any contribution arising at the element 'interfaces'. Alternatively, such interelement contributions must be of a kind that decrease continuously with the fineness of element subdivisions. The class of functions satisfying the first conditions shall be called *conforming* whilst the ones which satisfy only the second one are named *non-conforming* (but usually admissible).

In general it is quite easy to specify the 'conformity' conditions. If the integrand contains mth derivatives of the unknown functions ϕ then the shape functions **N** have to be such that the function itself and its derivatives up to the order $m - 1$ are required to be continuous (C^{m-1} continuity).[1]

In practice it is difficult to define conforming functions in a piecewise manner for any order of m greater than one and, because of this, many 'non-conforming' elements have originated in the past with the hope, sometimes proved *a posteriori*, that admissibility is achieved. The question of establishing admissibility is a difficult one and much work on this area is highly mathematical and not easy to interpret.[14,15] Simple tests of admissibility have however been devised and it is essential to subject any new non-conforming element to such an examination.[16,17]

In Table 1.2 we show these two main directives on which finite element approximation is based. There is however an intermediate position where interelement contributions can be evaluated without the imposition of full conformity. This arises either if the derivatives of **N** occur in a linear form in the integrals and continuity can be relaxed by one further order[18,19] or where in the basic formulation interface contributions are specifically inserted. The latter is the position with certain hybrid formulations[20,21] or Lagrange multiplier forms which specifically impose conformity as a constraint.[22,23] In this simple exposition we shall not be further concerned with these special situations, treating the *conforming* formulation as *standard* and the non-conforming one as a special variant of it.

A further condition which has to be imposed on shape function is that of 'completeness', i.e. the requirement that in the limit, as the element size decreases indefinitely, the combination of trial functions should exactly reproduce the exact solution. This condition is simple to satisfy[1,17] if

Table 1.2 Nature of trial functions used in finite element methods
(Ω—problem domain; Γ—boundary; I—element interface)

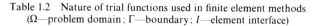

Conforming (square integrable) trial functions in Ω	Integrable trial functions $\int_{I^e} (\ ) \, dI$ with interface terms	Non-conforming trial functions (non-integrable)

$$\int_{\Omega} = \sum \int_{\Omega^e}$$

standard element with no interface terms — 'Jump' contributions derived mathematically — Lagrangian interface constraints — Engineering 'intuition' suggest functions

Exact integrals

Numerically sufficient integration

'Interweaving' or Taylor expansion forms guarantee convergence FINITE DIFFERENCE BASIS

Substitute non-conforming trial functions derived as least square approximation of conforming functions

polynomial expressions are used in each element such that the complete *m*th order of polynomial is present, when *m*th order derivatives exist in the integral of approximations.

To demonstrate a few simple 'shape' functions we show in Figures 1.1 and 1.2 some piecewise defined shapes in one- and two-dimensional domains

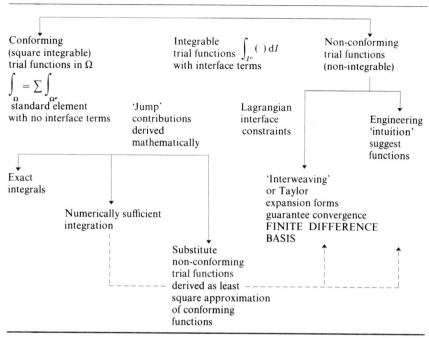

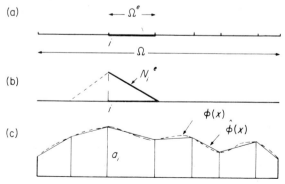

Figure 1.1 One-dimensional, simple, localized trial functions of C^0 continuity. (a) Domain subdivided into elements. (b) Localized trial function for parameter $a_i = \phi_i$. (c) Approximation to an arbitrary function $\phi(x)$

Finite Elements in Fluids

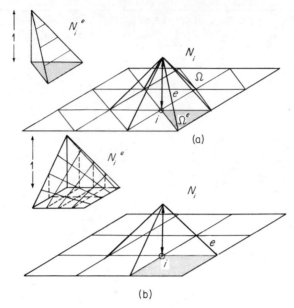

Figure 1.2   Two-dimensional localized trial functions of C^0 continuity. (a) Triangular element (linear expansion). (b) Rectangular elements (bi-linear expansion)

in which, to ensure C^0 conformity, the system parameter a_i takes on simply the value of the unknown function of certain points (often referred to as nodes) which are common to more than one 'element'. With this device a simple repeatable formula can be assigned to define N_i within any element.

It will be immediately recognized that as the parameter a_i influences the value of ϕ only in elements adjacent to a 'node' i its contribution to the integrals will be limited to elements containing that node—hence the 'banded' feature of approximating equations already referred to.

It is of interest to note that such piecewise defined functions, which to many form the essence of the finite element method, were used for the first time in 1943 by Courant[24] despite the fact that integral approximation procedures in their general form are much older.

Today many complex forms of shape function definition exist, mostly developed in the last decade,[1] which are capable of being piecewise defined and giving high orders of approximation. Some such functions are in fact defined, not in the simple original coordinate system in which the problem is given, but with a suitable transformation referred to curvilinear coordinates, by means of which most complex shapes of regions can be subdivided. Figure 1.3 shows some such elements of an 'isoparametric' kind[1,25] much used in practice. In all C^0 continuity only is imposed and relatively simple formulation suffices.

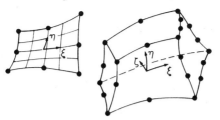

Figure 1.3 Some more elaborate elements with curvilinear coordinates

1.5.2 Particular example

To illustrate the process of discretization which by this time must appear somewhat abstract to the reader we shall return to the specific example given by Equation 1.6 and its associated boundary conditions, Equation 1.7.

Assuming that the potential ϕ (here a *scalar* quantity) can be approximated as

$$\hat{\phi} = \sum N_i a_i = \mathbf{N}\mathbf{a} \tag{1.28}$$

in which both N_i and a_i are scalars and a_i is identified with nodal values of ϕ we shall first use the variational principle of Equation (1.8).

Substituting Equation 1.28 into Equation 1.8 and differentiating with respect to a parameter a_j gives a set of equations required for stationarity as

$$F_j \equiv \frac{\partial \hat{\Pi}}{\partial a_j} = \frac{\partial}{\partial a_j}\left[\iint_\Omega \left\{ \tfrac{1}{2}k\left(\frac{\partial}{\partial x}(\sum N_i a_i)\right)^2 + \tfrac{1}{2}k\left(\frac{\partial}{\partial y}\sum N^i a_i\right)^2 - Q\sum N_i a_i\right\} \mathrm{d}x\,\mathrm{d}y\right.$$

$$\left. - \int_{\Gamma_2} q \sum N_i a_i \, \mathrm{d}\Gamma \right.$$

$$= \iint \left[k\frac{\partial N_j}{\partial x}\sum \frac{\partial N_i}{\partial x}a_i + k\frac{\partial N_j}{\partial y}\sum \frac{\partial N_i}{\partial y}a_i - QN_j \right]\mathrm{d}x\,\mathrm{d}y$$

$$- \int_{\Gamma_2} qN_j \, \mathrm{d}\Gamma \tag{1.29}$$

The whole equation system can be written as

$$\frac{\partial \hat{\Pi}}{\partial \mathbf{a}} = \mathbf{K}\mathbf{a} + \mathbf{P} = 0 \tag{1.30}$$

with

$$K_{ij} = \iint_\Omega k\left(\frac{\partial N_i}{\partial x}\cdot\frac{\partial N_j}{\partial x} + \frac{\partial N_i}{\partial y}\cdot\frac{\partial N_j}{\partial y}\right)\mathrm{d}x\,\mathrm{d}y$$

$$P_j = -\iint_\Omega QN_j\,\mathrm{d}x\,\mathrm{d}y - \int_{\Gamma_2} qN_j\,\mathrm{d}\Gamma \tag{1.31}$$

With trial function assumed piecewise element by element it is simple to evaluate the integrals for each element obtaining their contributions K^e_{ij} and P^e_j and obtaining the final equation by simple summation over all elements

$$K_{ij} = \sum K^e_{ij}$$
$$P_j = \sum P^e_j \qquad\qquad (1.32)$$

Alternative forms of approximation can be derived by the reader using some weighting procedures described. It can be shown that in this linear case (i.e. in which k and Q are functions of position only) identical approximation will be available by application of the Galerkin weighting but that other approximations will arise from use of alternative procedures. He will find that, for instance, application of least square processes Equation 1.22 will result in second derivatives being present and will need C^1 continuity trial functions with subsequent difficulties of determining such functions. He will however observe that the Galerkin process is available for non-linear problems where the simple form of the variational principle is no longer applicable.

Using the approximation defined in Equations 1.30–1.32 it is of interest to insert a particular shape function and obtain in detail a typical discretized equation for a parameter j. Let us consider a typical internal node $j = 0$ on a regular mesh of triangular elements as shown in Figure 1.4(a) in which a

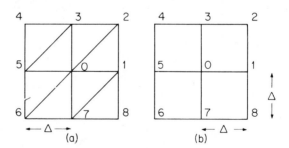

Figure 1.4   Regular triangle and square subdivisions

linear interpolation is used. Assuming that k and Q are constants and that the boundary does not occur in the vicinity, the contributions of all elements are found and coefficients K^e_{01}, K^e_{02} etc. evaluated. After assembly the typical equation becomes, after some algebraic manipulation, simply

$$k(\phi_1 + \phi_3 + \phi_5 + \phi_7 - 4\phi_0) + Q\Delta^2 = 0 \qquad\qquad (1.33)$$

The reader will recognize this as the standard finite difference equation obtained by direct, point, differencing of Equation 1.6 and he may well enquire what advantage has been gained. Obviously, the numerical answers in this case are going to remain the same (at least if boundary of the type Γ_2 does not occur). Immediately, however, it is important to point out that if k varied discontinuously between elements (such as may be the case at interfaces between two regions of different impermeability) the finite element method would have yielded, in one operation, answers which direct finite difference procedures would tackle only by the introduction of additional constraints and interface conditions. Further the variational form allows the gradient boundary condition to be incorporated directly.

Pursuing the problem further we try a rectangular element with a bilinear interpolation of ϕ as in Figure 1.4(b). The resulting finite element equation now becomes

$$k[\phi_1 + \cdots + \phi_8] - 8k\phi_0 + 6Q\Delta^2 = 0 \qquad (1.34)$$

a form substantially different from the standard finite difference equation which, although convergent to the same order of approximation, reduces the truncation error. Again the same comments can be made as before regarding the advantages of the finite element approximation.

In recent years much progress has been made in the finite difference methodology; in particular, integral forms including variational principles have been used as the basis of approximation in which the differences are only applicable to the differentials occurring in the integrals.[26,27,28,29,30,31,32] Comparison of such processes with finite element methods has been made by Pian[33] showing some detail of the problem discussed above. Such approaches eliminate some of the drawbacks of the finite difference procedures and indeed bring it close to the finite element process, as will be shown in the next section. However the difficulty of increasing the order of approximation or of using irregular meshes still preserves the advantages of finite element processes. Subsequent chapters will show many examples.

1.6 Some aspects of non-conformity—completing the circle to finite differences

As mentioned before, many non-conforming elements have been implemented in practice and convergence proofs obtained. Indeed very often these non-conforming elements have proved to produce results of higher accuracy than corresponding conforming ones. What is the reason for this? Is it desirable on occasion to introduce non-conformity deliberately to produce better results?

To answer these questions it is of interest to consider the terms on which the performance of an element is based. It is found by mathematical reasoning that the order of convergence of a particular element is dependent only on the complete polynomial terms which occur in the expansions.[17] This

Finite Elements in Fluids

indeed may be anticipated by considering the remainder terms in a local Taylor expansion of the unknown functions near the vicinity of a point of the domain.

In order to introduce conformity it is often found that either incomplete polynomial expansions are used, as for instance in the bilinear rectangle of Figure 1.2(b) in which only one quadratic term (x, y) occurs in addition to a complete linear expansion, or else, as in C^1 class elements,[1] singularities or rational fractions are introduced in addition to ordinary polynomial terms within the element. It is often the existence of such terms which causes the performance of an element to deteriorate and some means of eliminating these should perhaps be sought.

One answer to this problem has been recently supplied by Irons and Razzaque[34] who introduce the concept of *substitute shape functions*. The essential idea is to replace the original, conforming, shape which contains superfluous high order terms of expansion or singularities by another function which is a polynomial expansion complete to a certain order and which, in the least square sense, represents the best fit to the original shape function. Thus if N is the original function and $\overline{N}$ its polynomial substitute of the above kind,

$$\overline{N} = b_1 + b_2 x + b_3 y + \cdots \tag{1.35}$$

we determine the coefficients b by minimizing within each element

$$\int_{\Omega^e} (N - \overline{N})^2 \, d\Omega \tag{1.36}$$

with respect to these parameters. This results in a set of equations

$$\int_{\Omega^e} (N - \overline{N}) \frac{\partial \overline{N}}{\partial b_i} \, d\Omega \tag{1.37}$$

from which b_i are readily found. Clearly in general $\overline{N}$ will not be 'conforming'.

Convergence of elements derived on such substitute bases can be argued from the fact that in the limit (discounting any singularities present) the combination of either N or $\overline{N}$ is capable of representing a simple polynomial Taylor expansion in an identical manner.

The improvement of the performance of elements derived on this basis for some C^1 continuity problems in the context of plate bending has been demonstrated in Reference 35 and, indeed, in other situations a similar improvement is expected.

An apparently alternative path to the improvement of element performance has been recently demonstrated, in the context of numerically integrated elements, by reducing (rather than increasing) the order of numerical integration.[36] It is easy to show that one of the reasons for the success of this process is in fact its identity with the use of substitute shape functions. To show that it is of interest to record some properties of Gauss integration points used in numerical integration. Thus if n Gauss sampling points are used in an integration domain $-1 < x < 1$ then

(a) A polynomial of degree $2n - 1$ is integrated exactly.
(b) The n Gauss points define uniquely a polynomial of degree $n - 1$ which is the least square approximation of any polynomial of degree n which has the same sampling values.

Further we can observe that if $\bar{N}$ represents a least square approximation to N then $\partial\bar{N}/\partial x$ is also a least square approximation to $\partial N/\partial x$ etc. for all derivatives.

To illustrate these properties observe that any parabolic curve N^i in x direction is represented in a least square approximation by a straight line $\bar{N}$ passing through the two Gauss points—further the values of $\partial \dot{N}^i/\partial x$ and $\partial N/\partial x$ obtained at a single Gauss point sampling are identical, as shown in Figure 1.5.

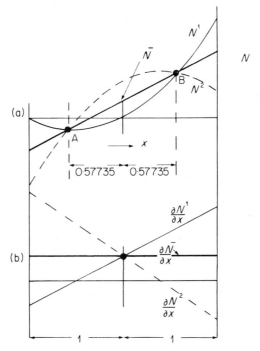

Figure 1.5 A property of Gauss points. (a) Two Gauss points define a straight line, $\bar{N}$, which is a least square approximation of any parabola N^i passing through these points. (b) $\partial\bar{N}/\partial x$ is a (constant) least square approximation to $\partial N^i/\partial x$

Finite Elements in Fluids

In practical applications one-dimensional domains are of little interest; but in two or three dimensions we observe immediately that for a bilinear expansion of Figure 1.2(b) the effects of sampling N or $\partial N/\partial x$ at one central Gauss point is equivalent to passing a least square substitute linear expansion $\bar{N}$ as shown in Figure 1.6.

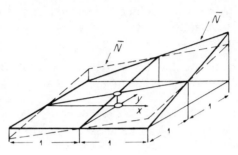

Figure 1.6 One Gauss point in a two-dimensional square samples correctly $\bar{N}$ and $\partial \bar{N}/\partial x$, $\partial N/\partial y$ which are least square approximations by a linear expansion to a bilinear function N

In second order rectangular elements used frequently in finite element analysis, Figure 1.7, terms such as $x^2 y$, xy^2 arise, which give first derivatives which vary parabolically in one direction. The effect of using a 2×2 Gauss sampling is to approximate to such derivatives by a bilinear expansion which

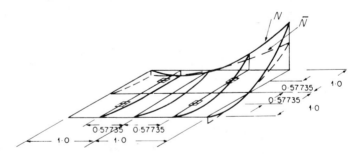

Figure 1.7 Four Gauss points in a two-dimensional square sample correctly $\bar{N}$ and its derivatives where $\bar{N}$ is a least square approximation by a parabolic expansion to a bi-parabolic function N

approximates these terms in a least square manner (thus approximating to the original shape function by a complete second order expansion). The success of 2×2 Gauss integration achieved in many situations is undoubtedly due to this fact. Table 1.3 shows a typical structural application in which dramatic improvement of results occurs by reduction of integration. In the context of fluid mechanics problems similar improvements of approximation again occur.

Table 1.3 Central deflections of a square plate under lateral uniform load.[1] Solution using four 'parabolic' three-dimensional elements with reduced integration (δ_c is the exact solution; L/t is the span/thickness ratio)

		$L/t = 200$	$L/t = 10$
$3 \times 3 \times 3$ Gauss point	$\dfrac{\delta}{\delta_c} =$	0·60	0·85
$2 \times 2 \times 2$ Gauss point	$\dfrac{\delta}{\delta_c} =$	1·00	0·98

The above remarks show that in many situations an improvement in the results is achieved by introduction of admissible, non-conforming, shape functions. At this stage it is of interest to examine the finite difference approximation and to show that these are in fact simple applications of such non-conforming trial function assumptions.

Consider for instance a 'direct' finite difference approximation to an ordinary differential equation

$$\frac{d^2\phi}{dx^2} + Q = 0 \tag{1.38}$$

The standard 'local' approximation to the second derivative in the vicinity of point, i.e.

$$\frac{d^2\phi}{dx^2} = \frac{1}{\Delta^2}(\phi_{n+1} - 2\phi_n + \phi_{n-1}) \tag{1.39}$$

is in fact identical to the choice of a trial function which is a parabola fitting the three consecutive values of ϕ. The governing equation well known in finite differences

$$\phi_{n+1} - 2\phi_n + \phi_{n-1} - \Delta^2 Q = 0 \tag{1.40}$$

can be obtained by an integral approximation of the form given by Equation 1.15 with a weighting function shown in Figure 1.8.

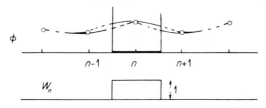

Figure 1.8 Equivalence of finite difference approximation to $d^2\phi/dx^2$ with a collocation finite element approach using discontinuous (non-conforming) trial functions

Clearly the shape functions chosen here are non-conforming and show discontinuities between successive elements. Convergence is dependent only on the 'interweaving' nature of those which guarantee that in the limit the discontinuity disappears.

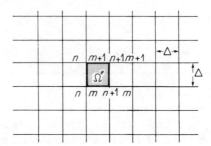

Figure 1.9 An equivalence of a finite difference approximation for first derivative in Ω^e to one point numerical sampling of bilinear expansion of Figure 1.8

Another much used finite difference approximation is presented in Figure 1.9 where, say in a variationally formulated problem, an expression for a gradient is written as

$$\frac{\partial \phi}{\partial x} \simeq \frac{1}{2\Delta}[\phi_{n+1,m+1} + \phi_{n+1,m} - \phi_{n,m+1} - \phi_{m,n}] \tag{1.41}$$

It is immediately evident that this is precisely the value obtainable by use of the substitute shape function in Figure 1.6 (or simply one point integration) and indeed identical approximation will result.

Pursuing the line of thought indicates that all finite difference processes can be considered as special cases of the finite element process with non-conforming, but usually admissible, shape function assumptions.

The success of finite difference methods is indeed dependent on the convergence of the trial function approximation and finite differences can be considered a particular case of the finite element process though originated in a different manner. It is more than likely that the future optimal methods of numerical discretization can borrow from the successes of both procedures. The finite element methodology based on irregular subdivision of elements and often a variational formulation frees the standard finite difference analyst from his shackles of regular mesh subdivision. The finite element method may well make greater use of non-conforming assumptions for trial functions implicit in the finite difference approximation. One such interesting 'marriage' was indicated recently by Utku[37] where an interweaving mesh is used on an irregular basis in two dimensions by passing a least square quadratic surface fit of a local expansion which yields second derivatives in an 'element' association with the locality as shown in Figure 1.10.

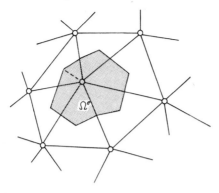

Figure 1.10 Non-conforming but interweaving and admissible expansion for ϕ in Ω^e by imposing a local parabolic expansion for ϕ_i in terms of a least square fit surface with neighbouring points (see Reference 37)

Questions such as the adequacy of 'interweaving' required to obtain admissibility of such non-conforming shape functions remain yet to be answered but, as mentioned before, pragmatic tests exist to judge whether convergence will be obtained. Problems such as instability of equation systems derived from certain function assumptions (e.g. for certain subdivisions using the one point integration rule of Figure 1.6 or of equivalent finite difference model of Equation 1.41) still need to be further investigated. Nevertheless, today it can be stated that the finite element methodology represents a very considerable generalization of the finite difference ideas and hence opens the way to solution processes which are at least as efficient. At best it opens new ways to problems which previously defied analysis.

Subsequent chapters will doubtless alert the reader to some possibilities offered.

References

1. M. J. Turner, R. W. Clough, H. C. Martin and L. J. Topp, 'Stiffness and deflection analysis of complex structures', *J. Aero Sci.*, **23**, 805–823 (1956).
2. R. W. Clough, 'The finite element in plane stress analysis', *Proc. 2nd ASCE Conf. on Electronic Computation, Pittsburgh, Pa., Sept. 1960.*
3. S. H. Crandall, *Engineering Analysis*, McGraw-Hill, New York, 1956.
4. B. A. Finlayson, *The Method of Weighted Residual and Variational Principles*, Academic Press, New York, 1972.
5. J. T. Oden, 'Finite element models of non-linear operator equations', *Proc. 3rd Conf. on Matrix Meth. in Struct. Mech., Wright-Patterson AFB, Ohio, 1971.*
6. E. Tonti, 'Variational formulation of non-linear differential equations', *Bull. Acad. Roy. Belgique*, Series 5, **55**, 139–165, 262–278 (1969).
7. R. S. Sandhu and K. S. Pister, 'Variational principles for boundary value and initial value problems in continuum mechanics', *Int. J. Solids Structures*, **7**, 639–654 (1971).

8. O. C. Zienkiewicz and C. Taylor, 'Weighted residual process in finite element method with particular reference to some transient and coupled problems', *Lectures on Finite Element Methods in Continuum Mechanics Proc. NATO Symp. Lisbon* (ed. J. T. Oden and E. R. A. Oliveira), Alabama Press, 1973.
9. O. C. Zienkiewicz, 'Constrained variational principles and penalty function methods in finite element analysis', *Conference on the Numerical Solution of differential equations, Univ. of Dundee, July 1973; Lecture Notes in Mathematics*, Springer Verlag, Berlin, 1974.
10. A. M. Arthurs, *Complementary Variational Principles*, Clarendon Press, Oxford, 1970.
11. P. P. Lynn and S. K. Arya, 'Use of the least squares criterion in finite element formulation', *Int. J. Num. Meth. Eng.*, **6**, 75–88 (1973).
12. O. C. Zienkiewicz, D. R. J. Owen and K. N. Lee, 'Least square finite element for elasto-static problems, use of reduced integration', *Int. J. Num. Meth. Eng.*, **8**, 341–358 (1974).
13. O. C. Zienkiewicz and P. N. Godbole, 'Flow of plastic and visco-plastic solids with special reference to extrusion and forming processes', *Int. J. Num. Meth. Eng.*, **8**, 3–16 (1974).
14. E. R. A. Oliveira, 'Theoretical foundations of the finite element method', *Int. J. Solids Structures*, **4**, 929–952 (1968).
15. P. C. Ciarlet, 'Conforming and non-conforming finite element methods for solving the plate problem', *Conference on the Numerical Solution of differential equations, University of Dundee, July 1973; Lecture Notes in Mathematics*, Springer Verlag, Berlin, 1974.
16. G. P. Bazeley, Y. K. Cheung, B. M. Irons and O. C. Zienkiewicz, 'Triangular elements in bending-conforming and non-conforming solutions', *Proc. Conf. Matrix Meth. Struct. Mech., Wright-Patterson AFB, Ohio, 1965.*
17. G. Strang and G. J. Fix, *An Analysis of the Finite Element Method*, Prentice-Hall, 1973.
18. S. Nemat-Nasser and K. N. Lee, 'Finite element formulations for elastic plates by general variational statements with discontinuous fields', *Report No. 39*, The Technical University of Denmark, 1973.
19. S. Nemat-Nasser and K. N. Lee, 'Applications of general variational methods with discontinuous fields to bending, buckling and vibration of beams', *Comp. Methods in Appl. Mech. Eng.*, **2**, 1, 33–41 (1973).
20. T. H. H. Pian and Pin Tong, 'Finite element methods in continuum mechanics', *Advances in Applied Mechanics*, Vol. 12, Academic Press, New York, 1972.
21. P. Tong, 'New displacement hybrid finite element model for solid continua', *Int. J. Num. Methods Eng.*, **2**, 78–83 (1970).
22. J. W. Harvey and S. Kelsey, 'Triangular plate bending element with enforced compatibility', *AIAA J.*, **9**, 6, 1023–1026 (1971).
23. B. A. Szabo and C. T. Tsai, 'The quadratic programming approach to the finite element method', *Int. J. Num. Meth. Eng.*, **5**, 375–381 (1973).
24. R. Courant, 'Variational methods for the solution of problems of equilibrium and vibration', *Bull. Am. Math. Soc.*, **49**, 1–23 (1943).
25. O. C. Zienkiewicz, B. M. Irons, J. Ergatoudis, S. Ahmad and F. C. Scott, 'Isoparametric and associated element families for two- and three-dimensional analysis', in I. Holand and K. Bell (Eds.), *Finite element method in stress analysis*, Tapir, Trondheim, Norway, 1969.
26. R. Courant and P. Hilbert, *Methods of Math. Physics.*, Vol. 1, Interscience, New York, 1953.

27. G. E. Forsythe and W. R. Wasow, *Finite Difference Methods for Partial Differential Equations*, Section 20, Wiley, New York, 1960.
28. D. M. Young, Jr., 'Survey of numerical analysis', in John Todd (Ed.), *The numerical solution of elliptic and parabolic partial differential equations*, Addison-Wesley, 1972.
29. D. Greenspan, 'On approximating extremals of functionals—1. The method and example for boundary value problems', *Bull. Int. Comp. Centre*, **4**, 99–120 (1965).
30. W. C. Walton, Jr., 'Application of a general finite difference method for calculating bending deformations of solid plates', *NASA TN D-536* (1960).
31. H. G. Schaefer and W. L. Heard, Jr., 'Evaluation of an energy method using finite differences for determining thermal midplane stresses in plates', *NASA TN D-2439* (1964).
32. D. S. Griffin and R. B. Kellog, 'A numerical solution of axially symmetrical and plane elasticity problems', *Int. J. Solids and Structures*, **3**, 781–794 (1967).
33. T. H. H. Pian, 'Variational formulation of numerical methods in solid continua', *Proc. Symposium on Computer Aided Engineering, Univ. of Waterloo, Ontario, Canada, 1971*, (Ed. G. H. L. Gladwell).
34. B. M. Irons and A. Razzaque, 'Shape function formulations for elements other than displacement models', *Symp. on Variational Methods, Univ. of Southampton, 1972*, pp. 4/59–4/71.
35. A. Razzaque, 'Program for triangular bending element with derivative smoothing', *Int. J. Num. Methods Eng.*, **6**, 333–345 (1973).
36. O. C. Zienkiewicz, J. Too and R. L. Taylor, 'Reduced integration technique in general analysis of plates and shells', *Int. J. Num. Meth. Eng.*, **3**, 275–290 (1971).
37. R. A. Ney and S. Utku, 'An alternative for the finite element method', *Symp. Variational Methods, Univ. of Southampton, 1972*, Southampton University Press, p. 3/62.

Chapter 2

Viscous, Incompressible Flow with Special Reference to Non-Newtonian (Plastic) Fluids

O. C. Zienkiewicz and P. N. Godbole

2.1 Introduction

In this chapter we are concerned with the problems of incompressible viscous flow and their discretization by the finite element process. As attention will be given elsewhere to the formulation and solution of high velocity situations we shall, in the main, concentrate our examples on creeping type problems in which dynamic terms can be neglected. The approach given here is, however, completely general and forms the basis of dealing with most incompressible fluid mechanics problems and in principle is followed in later chapters.

The approach to discretization taken here is via direct integral statements available from the use of virtual work principles in a manner similar to that widely practised in solid mechanics. This has a double advantage over processes which start explicitly from the governing Navier–Stokes equations. In the first place tedious algebraic operations are avoided; in the second, a directly analogous treatment to that of solid mechanics becomes available and it is possible not only to gain a deeper insight into problems presented but also to use directly many of the available programs of solid mechanics.

Much practical need exists for creeping flow solutions for non-Newtonian fluids in which the viscosity is strain-rate dependent. All the formulations presented here are valid for such problems and indeed we shall demonstrate several applications of this, non-linear, kind.

Of particular interest in non-Newtonian flow are problems of metal plasticity when large deformation occurs and elastic strains are negligible. Such problems, which stand on the borderline of fluid and solid mechanics, occur frequently in all kinds of extrusion and forming processes and have much current interest.

2.2 Basic formulation for viscous flow

By contrast to the solid mechanics problem, where we are primarily concerned with static response and concentrate on a *displacement* of a material point, in fluid mechanics the main variable of interest is the *velocity* at a point in space. The Lagrangian description is thus predominant in solids while here we shall follow an Eulerian one. Nevertheless the similarity of general concepts is great and we shall therefore borrow heavily from the methodology used in solids.[1]

Let u, v, w describe the three, Cartesian components of the velocity $\mathbf{u}$ of a point of space with coordinates x, y, z. Further let $\mathbf{X}$ denote the body forces per unit volume in part due to external causes X_0 and in part representing the dynamic acceleration effects. Thus,

$$\mathbf{X} = \mathbf{X}_0 - \rho \mathbf{A} \tag{2.1}$$

in which $\mathbf{A}$ stands for the acceleration of a point in the fluid, and ρ is the density.

If equilibrium of the internal stresses $\boldsymbol{\sigma}$ and body forces is considered, we have precisely the same equation set as in solid mechanics problems, i.e.

$$\mathbf{L}^T \boldsymbol{\sigma} + \mathbf{X} = 0 \tag{2.2}$$

where $\mathbf{L}$ is a linear operator obtainable from the well known explicit equations:

$$\frac{\partial}{\partial x}\sigma_x + \frac{\partial}{\partial y}\sigma_{xy} + \frac{\partial}{\partial z}\sigma_{xz} + X = 0 \tag{2.3}$$

etc.

with X being the x component of $\mathbf{X}$ etc.

A major difference from solid mechanics is due however to the Eulerian description of motion which, even in steady state flow, results in an acceleration $\mathbf{A}$.

Consider the acceleration component in the direction x for a mass point which has a velocity $\mathbf{u}$ at a point of space x, y, z. Its velocity rate of change for a particle of fluid depends not only on rates of change of $\mathbf{u}$ with respect to time but also on the changes of position. Thus for the x component of acceleration we have

$$A_x = \frac{D}{dt}u = \frac{\partial u}{\partial t} + \frac{\partial u}{\partial x}\cdot\frac{dx}{dt} + \frac{\partial u}{\partial y}\cdot\frac{dy}{dt} + \frac{\partial u}{\partial z}\cdot\frac{dz}{dt} \tag{2.4}$$

As $dx/dt = u$ etc. we have a definition of the 'convective' acceleration operator

$$\frac{D}{dt} \equiv \frac{\partial}{\partial t} + u\frac{\partial}{\partial x} + v\frac{\partial}{\partial y} + w\frac{\partial}{\partial z} = \frac{\partial}{\partial t} + \mathbf{u}^T . \text{ grad } \mathbf{u} \tag{2.5}$$

Here lies the major difference from solid mechanics where, as the displacements are referred to a particle and not to an element of space, only the simple differentiation of displacement with respect to time suffices.

To complete the formulation in solid mechanics problems we introduce a definition of strain in terms of displacements and a constitutive law defining a stress–strain relation.[1] For fluids we shall proceed similarly. First a rate of deformation, $\dot{\boldsymbol{\varepsilon}}$, is defined in terms of velocities. Thus we write

$$\dot{\boldsymbol{\varepsilon}} = \mathbf{L}\mathbf{u} = [\dot{\varepsilon}_{xx}, \dot{\varepsilon}_{yy}, \dot{\varepsilon}_{zz}, \dot{\varepsilon}_{xy}, \dot{\varepsilon}_{yz}, \dot{\varepsilon}_{zx}]^{\mathrm{T}} \tag{2.6}$$

with $\dot{\varepsilon}_{xx} = \partial u/\partial x$ etc. defining the operator $\mathbf{L}$. If tensorial representation is preferred the equivalent definition is

$$\dot{\varepsilon}_{ij} = \frac{1}{2}\left(\frac{\partial u_i}{\partial x_j} + \frac{\partial u_j}{\partial x_i}\right) \qquad i = 1\text{–}3 \tag{2.6a}$$

The constitutive relationship for fluids is more complex than in the solid mechanics problem as, in general, stresses depend not only on rates of strain but also on the strain itself. We shall therefore restrict our attention here to *incompressible* flow for which the rate of volumetric straining is zero, i.e. with $\mathbf{M}^{\mathrm{T}} = [1, 1, 1, 0, 0, 0]$; we write

$$\dot{\varepsilon}_v \equiv \dot{\varepsilon}_{xx} + \dot{\varepsilon}_{yy} + \dot{\varepsilon}_{zz} \equiv \varepsilon_{ii} \equiv \operatorname{div}\mathbf{u} \equiv \mathbf{M}^{\mathrm{T}}\dot{\boldsymbol{\varepsilon}} = 0 \tag{2.7}$$

For such fluids the mean stress σ is not defined and has to be sought from equilibrium relations. Defining the mean stress σ, or the pressure p, as

$$\sigma \equiv -p \equiv \frac{\sigma_{xx} + \sigma_{yy} + \sigma_{zz}}{3} \tag{2.8}$$

we obtain the deviatoric portion of the stress, $\mathbf{S}$, as a function of strain rate $\dot{\boldsymbol{\varepsilon}}$ in a matrix notation as

$$\mathbf{S} \equiv \boldsymbol{\sigma} - \begin{Bmatrix} 1 \\ 1 \\ 1 \\ 0 \\ 0 \\ 0 \end{Bmatrix} \sigma \equiv \boldsymbol{\sigma} + \mathbf{M}p \equiv \mathbf{R}\boldsymbol{\sigma} = f(\dot{\boldsymbol{\varepsilon}}) \tag{2.9}$$

with

$$\mathbf{R} = \begin{bmatrix} \frac{2}{3} & -\frac{1}{3} & -\frac{1}{3} & 0 & 0 & 0 \\ & \frac{2}{3} & -\frac{1}{3} & 0 & 0 & 0 \\ & & \frac{2}{3} & 0 & 0 & 0 \\ & \text{SYMMETRICAL} & & 1 & 0 & 0 \\ & & & & 1 & 0 \\ & & & & & 1 \end{bmatrix}$$

For a linear fluid we can write relation (2.9) as

$$\mathbf{S} = \mathbf{D}\dot{\boldsymbol{\varepsilon}} \tag{2.10}$$

where $\mathbf{D}$ is a matrix of constants. In tensorial notation we can rewrite above as

$$S_{ij} \equiv \sigma_{ij} - \tfrac{1}{3}\sigma_{ii} = D_{ijkl}\dot{\varepsilon}_{kl} \tag{2.10a}$$

This is entirely analogous to the definition of behaviour of elastic solids which are incompressible with $\mathbf{D}$ playing the role of the matrix of elastic constants. For isotropic linear behaviour it can be readily shown that we can write

$$\mathbf{D} = \mu \begin{bmatrix} 2 & & & & & 0 \\ & 2 & & & & \\ & & 2 & & & \\ & & & 1 & & \\ & & & & 1 & \\ 0 & & & & & 1 \end{bmatrix} = \mu\mathbf{D}^0 \quad \text{OR} \quad D_{ijkl} = 2\mu \tag{2.11}$$

in which μ is known as viscosity and $\mathbf{D}^0$ is a diagonal matrix. In general μ will be a function of $\dot{\varepsilon}$ and the formulation that follows is applicable in this form to non-Newtonian (non-linear) fluids.

The viscous flow problem is now fully defined and we can formulate its approximate solution mathematically proceeding formally by Galerkin or other weighting method. Such formal approaches are given in Chapters 12–14. However, an examination of the equations governing the flow and of the boundary conditions, which specify either tractions or velocities at all external boundaries, permits us to adapt here the procedures used in solid mechanics. In particular the 'virtual work principle' can be applied with *virtual velocities* playing the part of the virtual displacements used in solid mechanics.

Indeed one can conclude that the solution of a viscous flow problem is *identical to the solution of an equivalent incompressible elastic problem in which displacements are replaced by velocities and the body forces described by Equation 2.1 are inserted.*

Thus all the techniques available in the literature for the solution of one problem are available to the other—applicability being direct if the fluid flow is sufficiently slow for acceleration effects to be disregarded.

2.3 Viscous flow—velocity as the basic unknown

2.3.1 Virtual work statements

The equilibrium statement, Equation 2.3, can be replaced by an equivalent virtual work statement requiring that for any virtual velocity and strain rate changes, $\delta\mathbf{u}$ and $\delta\dot{\boldsymbol{\varepsilon}}$ which are compatible, external and internal rates of work are identical for any bounded domain. Thus we can write

$$\int_\Omega \delta\dot{\boldsymbol{\varepsilon}}^T\boldsymbol{\sigma}\,d\Omega - \int_\Omega \delta\mathbf{u}^T\mathbf{X}\,d\Omega - \int_{\Gamma_T} \delta\mathbf{u}^T\overline{\mathbf{T}}\,d\Gamma = 0 \qquad (2.12)$$

for any flow domain Ω in which tractions $\overline{\mathbf{T}}$ are specified on boundary $\Gamma_{\overline{T}}$ and where $\delta\mathbf{u}$ is zero on boundary Γ_u where velocities are given. Thus, for compatibility we have, from Equation 2.6,

$$\delta\dot{\boldsymbol{\varepsilon}} = \mathbf{L}\,\delta\mathbf{u} \qquad \text{and} \qquad \delta\mathbf{u} = 0 \quad \text{on } \Gamma_u \qquad (2.13)$$

As $\boldsymbol{\sigma}$ is not uniquely defined by $\dot{\boldsymbol{\varepsilon}}$ (being indeterminate due to the undefined pressure p), an additional equation needs to be written to enforce the incompressibility. As for any pressure variation δp internal work is zero due to incompressibility we can write

$$\int_\Omega \delta p\,\dot{\varepsilon}\,d\Omega = 0 \qquad (2.14)$$

From Equations 2.9 and 2.10 we find that $\boldsymbol{\sigma} = \mu\mathbf{D}\dot{\boldsymbol{\varepsilon}} - \mathbf{M}p$ and, inserting this into (2.12) we have

$$\int_\Omega \delta\dot{\boldsymbol{\varepsilon}}^T\mu\mathbf{D}^0\dot{\boldsymbol{\varepsilon}}\,d\Omega + \int_\Omega \delta\varepsilon_v p\,d\Omega - \int_\Omega \delta\mathbf{u}^T\mathbf{X}\,d\Omega$$

$$- \int_{\Gamma_T} \delta\mathbf{u}^T\overline{\mathbf{T}}\,d\Gamma = 0 \qquad (2.15)$$

Observing that by (2.7) we have

$$\dot{\varepsilon}_v = \dot{\varepsilon}_{xx} + \dot{\varepsilon}_{yy} + \dot{\varepsilon}_{zz} \equiv \mathbf{M}^T\dot{\boldsymbol{\varepsilon}} \qquad (2.15a)$$

Equations 2.15 and 2.14 lead directly to a discretization. We have now several choices open. First, we can allow an unrestricted definition of the fields of $\mathbf{u}$ and p and proceed using both equations. Second, we can confine our attention to velocity fields which automatically satisfy incompressibility. In the latter case Equation 2.14, as well as the second term of Equation 2.15, disappears and the full approximation involves only velocity parameters. We shall explore both formulations in turn.

If μ and $\mathbf{X}$ are for the moment treated as functions of position only it will be noted that Equations 2.15 and 2.14 are in fact equivalent to a variational principle of making a functional

$$\Pi = \int_\Omega \tfrac{1}{2}\dot{\varepsilon}\mu\mathbf{D}^0\dot{\varepsilon}\,d\Omega - \int_\Omega \mathbf{u}^T\mathbf{X}\,d\Omega - \int_{\Gamma_T} \mathbf{u}^T\overline{\mathbf{T}}\,d\Gamma \qquad (2.16)$$

stationary with respect to variations of *u* and subject to the constraint

$$\dot\varepsilon_v = \mathbf{M}^{\mathrm T}\dot{\boldsymbol\varepsilon} = 0 \tag{2.16a}$$

i.e. requiring stationarity of

$$\bar\Pi = \Pi + \int_\Omega p\dot\varepsilon_v \, \mathrm{d}\Omega$$

The pressure, *p*, in Equation 2.14 plays here the role of a Lagrangian multiplier introduced to satisfy the constraint.

We make this remark not only because the reader will observe, in the next section, that the characteristic structure of the discretized equations will contain the usual drawbacks of Lagrange multiplier forms (with zero diagonal terms in the matrices) but also because other ways exist of enforcing the constraint (2.16a).

2.3.2   *Discretization with velocity and pressure fields*

We describe, in the usual manner, the displacement and pressure fields by trial functions as

$$\mathbf{u} = \sum N_i^u a_i^u = \mathbf{N}^u \mathbf{a}^u$$
$$p = \sum N_i^P a_i^P = \mathbf{N}^P \mathbf{a}^P \tag{2.17}$$

in which $\mathbf{N}^u$ and $\mathbf{N}^P$ are appropriate trial or shape functions, defined element by element. It will be observed from the nature of integrals involved that we require C^0 continuity for the velocity field but discontinuous trial functions can be used to describe the pressure field. Writing

$$\delta\boldsymbol\varepsilon = \mathbf{L}\,\delta\mathbf{u} = (\mathbf{L}\mathbf{N}^u)\,\delta\mathbf{a}^u; \; \delta\dot\varepsilon_v = \mathbf{M}^{\mathrm T}\delta\dot{\boldsymbol\varepsilon} = \mathbf{M}^{\mathrm T}\mathbf{L}\mathbf{N}^u\,\delta\mathbf{a}^u \tag{2.18}$$

and

$$\delta p = \mathbf{N}^P\,\delta\mathbf{a}^P$$

we observe that (2.15) and (2.14) are true for all variations $\delta\mathbf{a}$. Thus we have from Equation 2.15

$$\left[\int_\Omega (\mathbf{L}\mathbf{N}^u)^{\mathrm T}\mu\mathbf{D}^0(\mathbf{L}\mathbf{N}^u)\,\mathrm{d}\Omega\right]\mathbf{a}^u + \left[\int_\Omega (\mathbf{M}^{\mathrm T}\mathbf{L}\mathbf{N}^u)^{\mathrm T}\mathbf{N}^P\,\mathrm{d}\Omega\right]\mathbf{a}^P$$

$$- \int_\Omega \mathbf{N}^{u\,\mathrm T}\mathbf{X}\,\mathrm{d}\Omega - \int_{\Gamma_{\mathrm T}} \mathbf{N}^{u\,\mathrm T}\bar{\mathbf{T}}\,\mathrm{d}\Gamma = 0 \tag{2.19}$$

and from Equation 2.14

$$\left[\int_\Omega \mathbf{N}^{P\,\mathrm T}\mathbf{M}^{\mathrm T}\mathbf{L}\mathbf{N}^u\,\mathrm{d}\Omega\right]\mathbf{a}^u = 0 \tag{2.20}$$

For cases of slow steady state viscous flow where $\mathbf{X} = \mathbf{X}_0$ this results in a simple symmetric set of equations which can be written as

$$\begin{bmatrix} \mathbf{K}^u & \mathbf{K}^p \\ \mathbf{K}^{p\,T} & 0 \end{bmatrix} \begin{Bmatrix} \mathbf{a}^u \\ \mathbf{a}^p \end{Bmatrix} + \begin{Bmatrix} \mathbf{P}^u \\ 0 \end{Bmatrix} = 0 \qquad (2.21)$$

where

$$K_{ij}^u = \int_\Omega (LN_i^u)^T \mu \mathbf{D}_0 (LN_j^u) \, d\Omega$$

$$K_{ij}^p = \int_\Omega (\mathbf{M}^T LN_i^u)^T N_j^p \, d\Omega$$

$$P^u = \int_\Omega N_i^{u\,T} \mathbf{X}_0 \, d\Omega - \int_{\Gamma_T} N_i^{u\,T} \mathbf{\bar{T}} \, d\Gamma \qquad (2.22)$$

Indeed this formulation is almost identical to that used in linear, incompressible elasticity which has been derived by imposing constraints on an energy functional.[2] If μ is independent of velocity, Equation 2.21 represents a simple linear equation system but the formulation is valid, i.e. even when $\mu = \mu(\varepsilon)$.

When considering unsteady state then, the coefficients $\mathbf{a}$ are time dependent.

Returning to Equation 2.1 and the explicit expression for the acceleration of Equation 2.5 we note that we can write

$$\mathbf{X} = \mathbf{X}_0 + \rho \frac{\partial \mathbf{u}}{\partial t} + \rho \mathbf{J} \mathbf{u} \qquad (2.23)$$

where

$$\mathbf{J} = \begin{bmatrix} \dfrac{\partial u}{\partial x} & \dfrac{\partial u}{\partial y} & \dfrac{\partial u}{\partial z} \\[2mm] \dfrac{\partial v}{\partial x} & \dfrac{\partial v}{\partial y} & \dfrac{\partial v}{\partial z} \\[2mm] \dfrac{\partial w}{\partial x} & \dfrac{\partial w}{\partial y} & \dfrac{\partial w}{\partial z} \end{bmatrix}$$

The term $\mathbf{P}^u$ of Equation 2.21 now has to be amended by replacing $\mathbf{X}_0$ by $\mathbf{X}$ of Equation 2.23. We have now, in place of $\mathbf{P}$ a new quantity $\mathbf{\bar{P}}$,

$$\mathbf{\bar{P}}^u = \left[\int_\Omega N^{u\,T} \rho N^u \, d\Omega \right] \frac{d\mathbf{a}^u}{dt} + \left[\int_\Omega N^{u\,T} \mathbf{J} N^u \, d\Omega \right] \mathbf{a}^u$$

$$\equiv \mathbf{M}^u \frac{d\mathbf{a}^u}{dt} + \mathbf{\bar{K}}^u \mathbf{a}^u \qquad (2.24)$$

The discretized Equation 2.21 becomes now

$$\begin{bmatrix} (\mathbf{K}^u + \overline{\mathbf{K}}^u), \mathbf{K}^p \\ \mathbf{K}^{pT}, \quad 0 \end{bmatrix} \begin{Bmatrix} \mathbf{a}^u \\ \mathbf{a}^p \end{Bmatrix} + \begin{bmatrix} M^u & 0 \\ 0 & 0 \end{bmatrix} \frac{d}{dt} \begin{Bmatrix} \mathbf{a}^u \\ \mathbf{a}^p \end{Bmatrix} + \begin{Bmatrix} \mathbf{P}_0 \\ 0 \end{Bmatrix} = 0 \qquad (2.25)$$

In the above the transient case can be solved by time stepping procedures but even in steady state the form of matrix equations obtained by omitting the second transient term is *non-linear* and *non-symmetric*.

The matrix $\overline{\mathbf{K}}^u$ depends on the current velocity and its form is non-symmetric. This presents difficulties in solving viscous flow problems with appreciable inertia and several alternative procedures have been used.[3,4] In some, the non-linear and non-symmetric matrix $\overline{\mathbf{K}}^u$ is taken care of by repeated iteration using only the symmetric part $\mathbf{K}^u$; in others, attempts at a direct solution of the non-linear equations have been made using a non-symmetric solution scheme. We discuss this matter further in Section 2.6.4.

2.3.3 Discretization using incompressible velocity fluids

The most usual procedure of describing incompressible velocities is by the use of stream function in two-dimensional problems or by the introduction of a vector potential in three dimensions.

Thus, if we confine our attention to plane flow, with u and v velocity components in x and y directions, we can write

$$\begin{aligned} u &= -\frac{\partial \psi}{\partial y} \\ & \qquad\qquad \text{OR} \qquad \mathbf{u} = \hat{\mathbf{L}}\psi \qquad (2.26) \\ v &= \frac{\partial \psi}{\partial x} \end{aligned}$$

where ψ is the stream function.
It is easily verified that

$$\dot{\varepsilon}_v \equiv \dot{\varepsilon}_{xx} + \dot{\varepsilon}_{yy} \equiv \frac{\partial u}{\partial x} + \frac{\partial v}{\partial y} \equiv 0 \qquad (2.27)$$

In an axisymmetric flow similarly we can write for radial and axial velocity components

$$u = -\frac{1}{r}\frac{\partial \psi}{\partial y} \qquad v = \frac{1}{r}\frac{\partial \psi}{\partial x} \qquad (2.28)$$

and once again incompressibility is obtained.

Finally in three-dimensional flow we can similarly define the velocity in terms of a vector potential with three components

$$\boldsymbol{\psi}^T = [\psi_x, \psi_y, \psi_z] \qquad (2.29)$$

as

$$\mathbf{u} = \text{curl}\,\boldsymbol{\psi} = \hat{\mathbf{L}}\boldsymbol{\psi} \tag{2.30}$$

where

$$\hat{\mathbf{L}} = \begin{bmatrix} \dfrac{\partial}{\partial z} & 0 & -\dfrac{\partial}{\partial y} \\[2ex] 0 & \dfrac{\partial}{\partial z} & -\dfrac{\partial}{\partial x} \\[2ex] \dfrac{\partial}{\partial y} & -\dfrac{\partial}{\partial x} & 0 \end{bmatrix}$$

Again it is easily verified that incompressibility is satisfied as

$$\varepsilon_v \equiv \frac{\partial u}{\partial x} + \frac{\partial v}{\partial y} + \frac{\partial w}{\partial z} \equiv 0 \tag{2.31}$$

(as $\varepsilon_v = \text{div}\,\mathbf{u}$ and $\text{div}\,\text{curl}\,\boldsymbol{\psi} \equiv 0$).

With velocities specified on the boundaries it is possible (if boundaries are simply connected) to determine the stream function and its normal gradient (or vector potential components) there to within an arbitrary constant. For discretization therefore we can assume an expansion for ψ

$$\boldsymbol{\psi} = \mathbf{N}_i \mathbf{a}_i = \mathbf{N}\mathbf{a} \tag{2.32}$$

and use the virtual work expression (2.15) with the second term dropped (as it now is identically zero), i.e.

$$\int_\Omega \delta\dot{\boldsymbol{\varepsilon}}\mu\mathbf{D}_0\dot{\boldsymbol{\varepsilon}}\,d\Omega - \int_\Omega \delta\mathbf{u}^\mathrm{T}\mathbf{X}\,d\Omega - \int_{\Gamma_T} \delta\mathbf{u}^\mathrm{T}\overline{\mathbf{T}}\,d\Gamma = 0 \tag{2.33}$$

Writing for all cases considered above

$$\mathbf{u} = \hat{\mathbf{L}}\boldsymbol{\psi} = \hat{\mathbf{L}}\mathbf{N}\mathbf{a}; \qquad \dot{\boldsymbol{\varepsilon}} = \mathbf{L}\mathbf{u} = \mathbf{L}\hat{\mathbf{L}}\mathbf{N}\mathbf{a} \tag{2.34}$$

with corresponding variations, we can immediately obtain the discretized form of equations from which $\mathbf{a}$ can be obtained as (as usual taking $\delta\mathbf{a}^\mathrm{T}$ outside in (2.33) after substitution and equating the multiplier to zero).

$$\left[\int_\Omega (\mathbf{L}\hat{\mathbf{L}}\mathbf{N})^\mathrm{T}\mu\mathbf{D}_0(\mathbf{L}\hat{\mathbf{L}}\mathbf{N})\,d\Omega\right]\mathbf{a} - \int_\Omega (\hat{\mathbf{L}}\mathbf{N})^\mathrm{T}\mathbf{b}\,d\Omega - \int_{\Gamma_T} (\hat{\mathbf{L}}\mathbf{N})^\mathrm{T}\overline{\mathbf{T}}\,d\Gamma = 0 \tag{2.35}$$

or the usual form

$$\mathbf{K}\mathbf{a} - \mathbf{P} = 0 \tag{2.36}$$

with expressions for K_{ij} and P_i apparent from (2.35).

Immediately an observation can be made that the shape functions N now require C^1 continuity as second derivatives operate on these in the integrals. While in two-dimensional problems the use of such functions presents little difficulty in three dimensions no satisfactory piecewise defined functions are available.

Confining our attention to the scalar stream function ψ defined for axisymmetric or plane problems it is readily seen that the same shape functions as used for plate bending analysis[1] are available. It is therefore possible to use *any* of the numerous plate functions for solution of viscous flow. Indeed, for the linear case of slow viscous flow equations, the whole formulation may be identified with plate bending equation and any standard plate bending programme adapted immediately.

While in the first type of formulation, Section 2.3.2, we have 'borrowed' heavily from previous methods used extensively in the solution of incompressible solids, in the second type (Section 2.3.3) we have introduced standard fluid mechanics concepts for enforcing incompressibility. These do not appear to have been used widely in solid mechanics and a 'reverse borrowing' is obviously possible. The stream function concept can be used directly in solid mechanics and only recently has such a development been put into practice.[5]

Once again inclusion of the dynamic term can be made, pursuing the process of modifying the **P** term—as described in the previous section.

The use of stream function introduces some drawbacks into many problems. Unless the velocities are entirely prescribed on all boundaries it is often impossible to establish, *a priori*, the values of stream functions on some positions of the boundary. This is particularly serious in multiply connected boundaries, such as are presented by flows around obstacles etc. To overcome these difficulties it is necessary to introduce additional constraints on the rate of boundary work. This, even in linear problems, presents serious difficulties and, when formulating general problems, considerable thought should be given to the nature of boundary conditions. In References 5 and 6 we discuss this matter in detail. In Chapter 12, stream function approach is employed in detailed solution of the Navier–Stokes equations.

2.4 Viscous flow—Equilibrium and mixed formulation

The need for enforcing incompressibility has presented some difficulties in the velocity type of formulation used in the previous section, these arise because the stresses are not completely defined by the strain rates (see Equation 2.9). On the other hand stresses uniquely define the strain rates and it is obviously possible, with advantage, to use the equivalents of equilibrium virtual work statements or of the 'mixed' formulations well known in solid mechanics. Possibilities here are enormous and have only been barely explored. We shall restrict ourselves here to a brief statement of the equilibrium formulation.

If the unknown function is the stress field σ which is chosen so as to exactly satisfy the equilibrium conditions then the virtual work done by 'compatible' strain rates must be zero, i.e.

$$\int_\Omega \delta\sigma^T \dot{\varepsilon}\, d\Omega - \int_{\Gamma_u} \delta\overline{T}^T \overline{u}\, d\Gamma = 0 \tag{2.37}$$

in which Γ_u is now the portion of the boundary on which velocities $u = \overline{u}$ are specified, and $\delta\overline{T}$ are the boundary tractions resulting from stresses $\delta\sigma$, i.e.

$$\delta\overline{T} = G\,\delta\sigma \tag{2.38}$$

where G is a suitable matrix of direction cosines of the normal to the surface. From Equations 2.9 to 2.11 the strain rates are defined in terms of stresses as

$$R\sigma = \mu D_0 \dot{\varepsilon} \tag{2.39}$$

or

$$\dot{\varepsilon} = \frac{1}{\mu}(D_0^{-1}R)\sigma = \frac{1}{\mu}C_0\sigma \tag{2.40}$$

in which

$$C_0 = \begin{bmatrix} \frac{1}{3} & -\frac{1}{6} & -\frac{1}{6} & 0 & 0 & 0 \\ & \frac{1}{3} & -\frac{1}{6} & 0 & 0 & 0 \\ & & \frac{1}{3} & 0 & 0 & 0 \\ \text{SYMMETRICAL} & & & 1 & 0 & 0 \\ & & & & 1 & 0 \\ & & & & & 1 \end{bmatrix}$$

To achieve an equilibrating field, stresses can be defined in terms of a stress function set Φ as

$$\sigma = E\Phi + \Lambda \tag{2.41}$$

where Λ is a particular solution which equilibrates the applied body forces and Φ is so constrained as to satisfy prescribed boundary tractions.

Use of stress function in context of solid mechanics has been pioneered by Veubeke and Zienkiewicz[7] with a subsequent development by Sander,[8] which in a sense is an application of certain special mixed formulations.

Details of the procedure are discussed elsewhere but some of the difficulties in the fluid mechanics context should now be noted.

First if the flow is not so slow that dynamic terms can be neglected these will appear in the inhomogeneous terms defining the stress field Λ in terms of velocities. As velocities can only be obtained by integration of the stress field, similar difficulties will arise as with those encountered in dynamics of solids when equilibrating forms are used. These difficulties can be overcome as shown by Taborrock[9] but for practical application the equilibrating formulation appears simple only in cases of slow viscous flow.

Further, in three-dimensional problems the use of the stress function necessitates again C^1 continuity to be introduced in appropriate expansions. This is nearly impossible to achieve in practice. Two-dimensional use of the Airy stress function is however practicable and useful.

Possibilities of using a 'mixed' formulation in which both stress and velocity field simultaneously appear have not yet been explored and present a fruitful field for research.

2.5 Some other solution possibilities: Penalty function method

In the preceding we have used simple virtual work or, which is equivalent, Galerkin formulation. Other forms of discretization are obviously possible and have been used in practice.

In some, following classical procedures of fluid dynamics, the governing equations are rewritten in terms of both stream function and vorticity. Direct approximation can then be used with Galerkin or other processes. Alternatives with the use of least square principle are possible—and can be applied directly to the equations in terms of all the variables.

A simple direct possibility can be used if we consider the analogy of fluid flow and incompressible elastic formulation. In the latter, to sidestep the difficulties involved due to incompressibility in a direct displacement formulation, the engineers have often used standard displacement programs with *near* incompressibility imposed. In terms of elastic constants this is equivalent to using a high Poisson ratio (say 0·49) in place of 0·5 for which a singularity arises. For simple finite element representation this usually leads to inaccurate results but with isoparametric parabolic elements good accuracy can be found with Poisson ratio as high as 0·49995—especially if 'reduced integration' of 2×2 Gauss points is used. This procedure is particularly simple as existing finite element programs can be employed.

We now show that this formulation is equivalent to the use of the constrained variational principle of Equation 2.16/2.16a and the introduction of the constraint via a penalty function approach.[10] In this approach we replace the functional of Equation 2.16 by one embodying the square of the constraint multiplied by a 'penalty', a large positive number α, and seek its stationarity. Thus we start with a functional

$$\overline{\Pi} = \int \tfrac{1}{2}\varepsilon\mu \mathbf{D}^0\varepsilon \, d\Omega - \int \mathbf{u}^\mathsf{T}\mathbf{X} \, d\Omega - \int \mathbf{u}^\mathsf{T}\overline{\mathbf{T}} \, d\Gamma + \alpha \int_\Omega \varepsilon_v^2 \, d\Omega \qquad (2.42)$$

Inserting expression (2.7), i.e.

$$\dot{\varepsilon}_v = \mathbf{M}^\mathsf{T}\dot{\varepsilon} \qquad (2.7)$$

we note that the new functional can be written as

$$\overline{\Pi} = \int_\Omega \tfrac{1}{2}\dot{\varepsilon}\overline{D}\dot{\varepsilon} \, d\Omega - \int_\Omega \mathbf{u}^\mathsf{T}\mathbf{X} \, d\Omega - \int_{\Gamma_\mathsf{T}} \mathbf{u}^\mathsf{T}\overline{\mathbf{T}} \, d\Gamma \qquad (2.43)$$

with

$$\mathbf{D} = \mu\mathbf{D}^0 + 2\alpha\mathbf{M}\mathbf{M}^\mathsf{T} \qquad (2.44)$$

Comparison of (2.43) and (2.44) with the solid mechanics problem immediately identifies the analogy of 2α and the bulk modulus of the solid. However, without any such physical reasoning we immediately note how a discretization can be directly achieved for the incompressible flow problem. Use of this procedure with isoparametric elements will be indicated in the next section.

2.6 Some illustrative linear-creeping flow applications

2.6.1 Entry flow—two and three dimensions

One of the first finite element solutions to such problems was achieved by Atkinson and coworkers with a stream function formulation.[11,12] In this situation the velocities have been assumed entirely specified on the external boundary and no prescribed tractions introduced. The singly connected boundary allows the stream function and its normal gradient to be determined from the velocity definitions (Equations 2.26 or 2.28) and no problems arise (an arbitrary constant in the value of ψ is obtained by specifying this at some point of the boundary).

In Figure 2.1 we show a simple solution of an axisymmetric entry flow region obtained by two alternative procedures. In the first, a stream function approximation is used with a rectangular, compatible element of four degrees of freedom at each corner (introduced by Bogner and coworkers).[13] (For details of this element see also Reference 1, p. 206.) In the second, the simple penalty function approach is used as described in the previous section and the variables are directly the velocities. Here a standard solid mechanics program[1] using parabolic isoparametric elements was employed with an effective Poisson ratio of 0·49995 and 2×2 Gauss rule. Both element subdivisions are noted in Figure 2.1 and velocity distributions are virtually not distinguishable. For comparison, results obtained by Lew and Fung[14] are given.

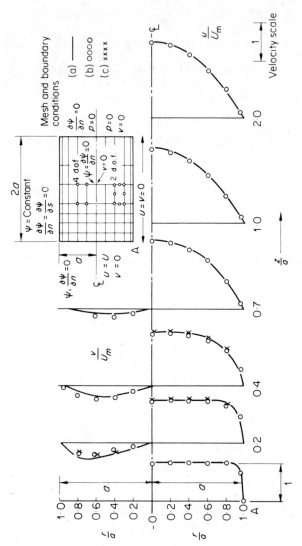

Figure 2.1 Velocity profile development in an axially symmetric entry flow. (a) by stream function approach (b) by $u - v$-penalty function approach ($v = 0.49995$). (c) Finite difference solution (Lew and Fung[14])

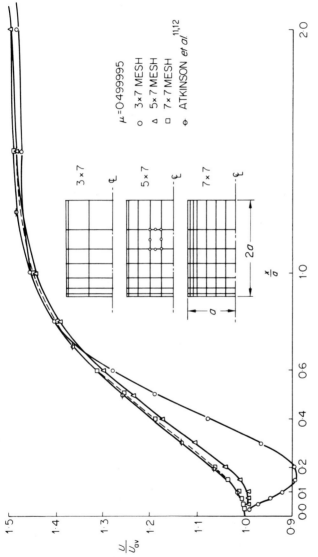

Figure 2.2 Effect of mesh subdivision on axial velocity development on the plane entry flow (on centre line): parabolic elements: penalty function approach

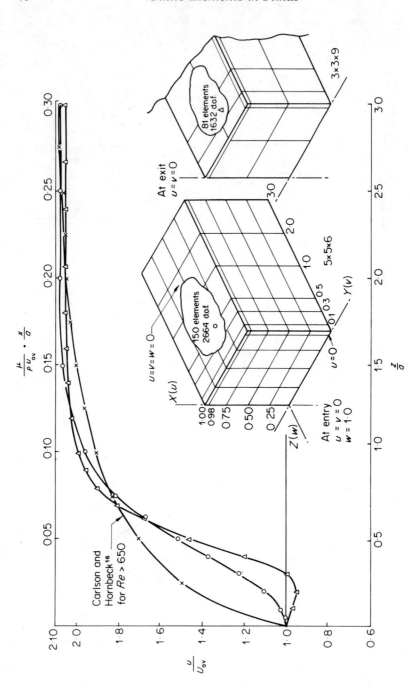

Figure 2.3 Axial velocity development at duct centre line (creeping flow)

To avoid difficulties introduced at the entry section where a velocity singularity is present at A, Figure 2.1, we have assumed a rapid but continuous transition of velocity near that point. Nevertheless, this near singularity introduces a considerable disturbance to the solution and quite fine subdivisions are needed for reasonable accuracy. In Figure 2.2 we show the effect of this subdivision on velocity development for a plane entry region. Here, once again, a standard displacement type program is used.

In a three-dimensional context, as already indicated, the use of stream function becomes impracticable. We show in the example of Figure 2.3 an entry velocity distribution in a rectangular conduit using again a penalty function approach—or indirectly a standard three-dimensional solid analysis program with $v = 0.49995$ and a $2 \times 2 \times 2$ Gauss integration.

The reader should note again similar errors to those shown in Figure 2.2 when a coarse mesh is used. In Figures 2.4 and 2.5 we show further

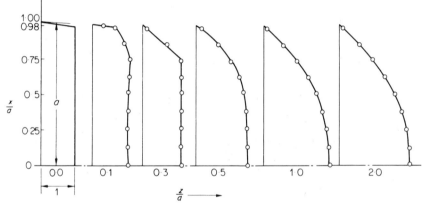

Figure 2.4 Axial velocity profiles on centre line section in Figure 2.3

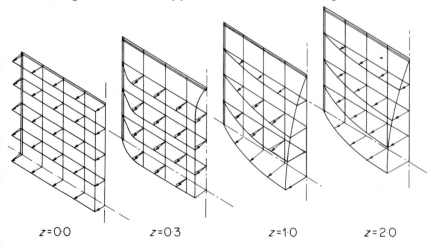

Figure 2.5 Three-dimensional entry flow problem (square duct): isometric view of velocity distribution

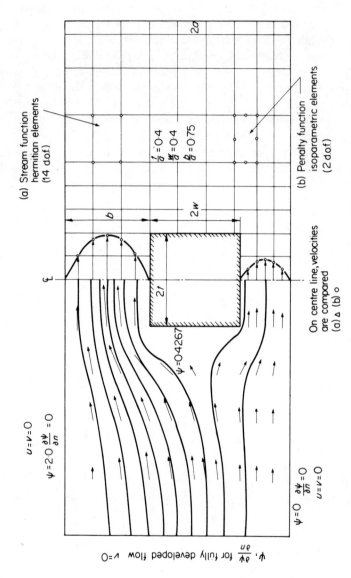

Figure 2.6 Two-dimensional flow with a square obstruction placed asymmetrically. (a) Stream function approach—streamlines shown. (b) Penalty function approach—velocity vectors shown

properties of the velocity distributions. No alternative creeping flow solution appears to be available and, for comparison, results of a higher Reynolds number investigation by Han and Carlson and coworkers[15,16] are shown.

2.6.2 Flow past asymmetrically and non-asymmetrically placed obstructions in parallel flow

This example is illustrated in Figure 2.6 in which once again both approaches used in the previous example have been employed. In Figure 2.6 (a) and (b) we give the velocity (streamline pattern obtained) while in Figure 2.7 wall

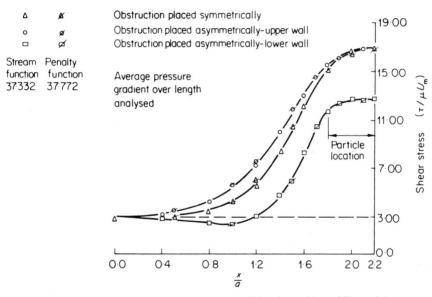

Figure 2.7 Shear stress distribution at the wall for the problem of Figure 2.6

shear stress distributions are compared. In the stream function approach the asymmetric placing of the obstacle necessitates a special treatment, discussed in Reference 6, in which two solutions for arbitrary values of stream function assumed at the inner boundary are superimposed. In the velocity penalty function formulation, no such difficulty is present and direct solution is obtained from both symmetric and asymmetric placing of the obstacle.

2.6.3 Use of triangular versus rectangular elements in stream function formulation

In the previous examples we have used simple rectangular elements for the stream function formulation as with these it is simple to satisfy C^1 continuity using polynomial expressions.[1,13] The problem of devising arbitrary triangular or quadrilateral elements with such a continuity is much more complex and in the text[1] the various difficulties and their solution are discussed. One of the most satisfactory triangular elements produced is one based on the original conforming triangle of Bazeley[17] using substitute smoothed (least square fit) shape function of cubic form.[18,19]

To test the efficiency of such elements a simple case of Poiseuille flow between parallel plates is considered (Figure 2.8). On two sections AA and

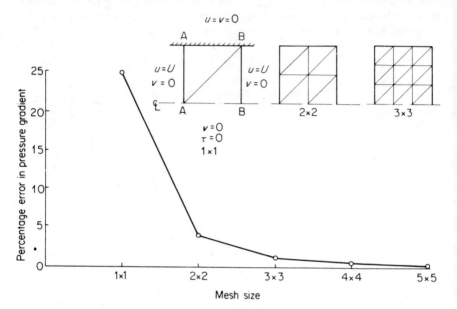

Figure 2.8   Convergence of triangular elements: plane Poiseuille flow: parabolic U prescribed on AA and BB

BB the velocity distribution is prescribed and the total pressure drop is compared with the exact solution. The rectangular elements for all subdivisions gave exact answers being totally conforming and having a complete cubic expansion available. With triangular elements an error arises and its decrease with fineness of subdivision is given in Figure 2.8. The convergence is approximately of order h^3.

Numerous direct solutions in terms of velocity and pressures are given for creeping flow problems in Reference 3.

2.7   Non-Newtonian flow

In non-Newtonian fluids the viscosity μ depends in same manner as the rate of straining $\dot{\varepsilon}$.

$$\mu = \mu(\dot{\varepsilon}) \tag{2.45}$$

Here using the formulations of Sections 2.2 to 2.5 we have (in the absence of dynamic terms) a discretized system of equations in the form

$$\mathbf{K}(\mu)\mathbf{a} - \mathbf{P} = 0 \qquad (2.46)$$

The matrix $\mathbf{K}$ is dependent on $\dot{\varepsilon}$ and hence on $\mathbf{a}$. A simple iterative procedure can now be adopted—and has been shown to converge quite rapidly even with substantially non-linear behaviour.

Assuming some value of $\mu = \mu^0$ the first solution is obtained

$$\mathbf{a}^1 = \mathbf{K}_0^{-1}\mathbf{P} \qquad (2.47)$$

and hence $\dot{\varepsilon}$ and new value of μ^1 at all points of the region is available to compute $\mathbf{K}_1$. The next approximation is

$$\mathbf{a}^2 = \mathbf{K}_1^{-1}\mathbf{P} \quad \text{etc.} \qquad (2.48)$$

leading to a standard iterative algorithm

$$\mathbf{a}^n = \mathbf{K}_{n-1}^{-1}\mathbf{P} \qquad (2.49)$$

It is usual to assume that viscosity is simply a function of the second strain rate invariant $\dot{\bar{\varepsilon}}$, i.e.

$$\dot{\bar{\varepsilon}} = 2\varepsilon_{ij}\varepsilon_{ij}$$

$$= 2\dot{\varepsilon}_{xx}^2 + 2\dot{\varepsilon}_{yy}^2 + 2\dot{\varepsilon}_{zz}^2 + (2\dot{\varepsilon}_{xy})^2 + (2\dot{\varepsilon}_{yz})^2 + (2\dot{\varepsilon}_{zx})^2 \qquad (2.50)$$

A frequently used expression is of a form

$$\mu = \mu_0(\dot{\bar{\varepsilon}})^{n-1} \qquad (2.51)$$

with index $n < 1$.

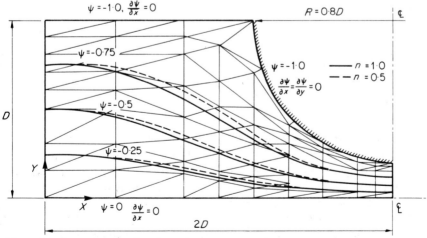

Figure 2.9 Streamlines for flow around a series of cylinders in a plane parallel flow: non-Newtonian fluid: streamlines

Some solutions for a fluid of this type are obtained by Palit and Fenner[20,21] using stream function formulation and 'incompatible' triangular elements. In Figure 2.9 solutions for two dimensional flow around a cylinder with $n = 1$ and $n = 0.5$ are compared for an assumed uniform entry flow velocity distribution at AA. The study illustrates well the local effects of non-linearity on velocity (streamlines) changes.

In this problem once again a stream function formulation is employed, and triangular elements with nine degrees of freedom are used to deal with the general boundary shape.

2.8 Plastic or visco-plastic behaviour of extruded metals—a case of non-Newtonian flow

A particularly interesting case of non-Newtonian creeping flow is that of a Bingham fluid or its generalization, the visco-plastic material. Such materials behave as solids exhibiting a zero rate of straining for stresses which are, in some measure, below a threshold or yield value. When this yield is exceeded, flow begins at a rate which is a function of the excess stress.

Let $F(\sigma) = 0$ represent this yield condition and we shall assume this if when $F(\sigma) < 0$ no flow occurs. Assuming further that the various components of strain rate are proportional to the gradients of F with respect to these (i.e. the so-called associated plasticity condition) we can describe with some generality the behaviour of the material by writing

$$\dot{\varepsilon} = \frac{1}{\bar{\mu}} \langle F^n \rangle \frac{\partial F}{\partial \sigma} \tag{2.52}$$

or in tensorial form

$$\dot{\varepsilon}_{ij} = \frac{1}{\bar{\mu}} \langle F^n \rangle \frac{\partial F}{\partial \sigma_{ij}} \tag{2.52a}$$

where $\langle \rangle$ means that

$$\langle F \rangle = 0 \quad \text{if } F < 0$$
$$\langle F \rangle = F \quad \text{if } F > 0 \tag{2.53}$$

and $\bar{\mu}$ is some 'viscosity' parameter.

It appears that we have here once again a case of viscous flow with a variable viscosity dependent now on the current stresses. It is however possible to reduce the problem to that discussed in the previous section where viscosity is a function of the strain rate (as in Equation 2.44).

To do this we shall find it convenient to use a tensorial notation and use the concept of deviatoric stress. Thus Equation 2.10 defining viscosity

$$\mathbf{S} = \mu \mathbf{D}_0 \dot{\varepsilon} \tag{2.54}$$

becomes in tensor notation

$$S_{ij} = \sigma_{ij} - \delta_{ij}\sigma_{ij} = 2\mu\dot{\varepsilon}_{ij} \qquad (2.54a)$$

Taking the yield conditions defined simply by the second invariant of S_{ij}, i.e. the so-called von Mises yield criterion

$$F = \sqrt{\tfrac{1}{2}S_{ij}S_{ij}} - \frac{1}{\sqrt{3}}\sigma_y \qquad (2.55)$$

in which σ_y is the yield stress in simple tension, we find that

$$\frac{\partial F}{\partial \sigma_{ij}} = \frac{\partial F}{\partial S_{ij}} = \frac{1}{2}\frac{1}{\sqrt{\tfrac{1}{2}S_{ij}S_{ij}}}S_{ij} \qquad (2.56)$$

and that we can write Equation 2.52a as

$$\dot{\varepsilon}_{ij} = \frac{1}{2\bar{\mu}}\left\langle \sqrt{\tfrac{1}{2}S_{ij}S_{ij}} - \frac{1}{\sqrt{3}}\sigma_y \right\rangle^{n} \cdot \frac{1}{\sqrt{\tfrac{1}{2}S_{ij}S_{ij}}}S_{ij} \qquad (2.57)$$

Comparison with the definition of viscosity in Equation 2.54a gives

$$\frac{1}{\mu} = \frac{1}{\bar{\mu}}\frac{\left\langle \sqrt{\tfrac{1}{2}S_{ij}S_{ij}} - \sigma_y \right\rangle^{n}}{\sqrt{\tfrac{1}{2}S_{ij}S_{ij}}} \qquad (2.58)$$

but we note that by (2.54a)

$$\sqrt{\tfrac{1}{2}S_{ij}S_{ij}} = \mu \qquad \sqrt{2\dot{\varepsilon}_{ij}\dot{\varepsilon}_{ij}} = \mu\dot{\bar{\varepsilon}} \qquad (2.59)$$

Inserting this in Equation 2.58 we see that μ can be obtained in terms of $\dot{\bar{\varepsilon}}$. For $n = 1$ it can easily be verified that

$$\mu = \frac{1}{\sqrt{3}}\frac{\sigma_y + \bar{\mu}\varepsilon}{\dot{\varepsilon}} \qquad (2.60)$$

and the general form identical with that discussed in previous section is obtained (Equation 2.45).

Computationally the expression (2.60) appears to present a difficulty as $\mu \to \infty$ with $\dot{\bar{\varepsilon}} \to 0$. However this problem is readily overcome by limiting the upper value of μ to some large number.

The visco-plastic model used is one of a class suggested by Perzyna[22] and is of quite wide applicability.[23] It is interesting to note that as the coefficient $\bar{\mu} \to 0$ the visco-plastic and plasticity formulation become identical. The solution procedure is therefore applicable to problems of both plastic and visco-plastic kind.

The solution of such plastic and visco-plastic problems is of great importance in the working (forming and extrusion) processes of metals and plastics and treatment proposed here is capable of providing simply such

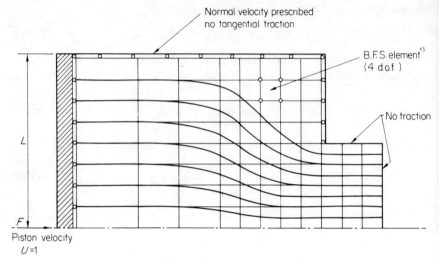

Figure 2.10 Element subdivision and resulting streamline pattern for the extrusion problem (plane strain): stream function formulation

solutions. In another paper[6] the authors outline some applications and illustrations of this process. Figures 2.10 to 2.12 show a solution to a simple plastic extrusion process through a die with 50 % reduction. With $\bar{\mu} = 0$ assumed the stress distribution and hence the forces required to sustain extrusion are independent of velocity of the piston U—and good comparison is obtained with results available for this case from simple slip line solutions. For different values of $\bar{\mu}$ both the piston force and the velocity pattern depend strongly on the value of the velocity of the piston. This dependence is shown in Figure 2.11.

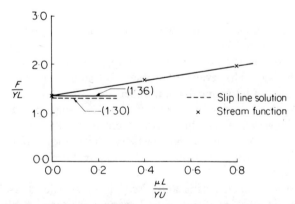

Figure 2.11 Variation of extrusion force in example of Figure 2.10 with $\bar{\mu}$ for a visco-plastic material ($U = 1$)

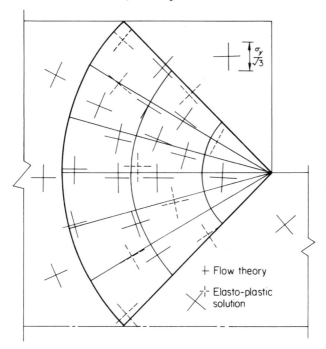

Figure 2.12 Directions, magnitudes of maximum shear stresses from flow and elasto-plastic solutions superimposed on the slip-line field: σ_y = uniaxial yield stress

2.9 Quasi-static transient solutions: Free surface problems

If the process of viscous flow is sufficiently slow for dynamic effects to be ignored we may still be faced with a transient (time variable) problem. If at a particular configuration with a boundary Γ_T on which traction is prescribed resulting velocities are not found to be entirely parallel to the surface, then clearly the configuration of the problem will change with time. A free surface is a particular example of such a problem which frequently arises in slow viscous processes such as indentation etc.

The treatment of such problems is relatively simple. Once the velocities on the boundary $\Gamma_{\bar{T}}$ are found, in the manner previously described, by a time-stepping extrapolation the free surface change can be predicted. Using a straightforward Eulerian prediction the change of position of a boundary in a direction **S** in a time increment can be predicted as

$$\Delta S = V_s \Delta t \qquad (2.61)$$

where V_s represents the corresponding velocity component.

Finite Elements in Fluids

Once the new position of the boundary is known a new solution can be readily obtained and the process continued.

In Figure 2.14 we show successive stages of an indentation process defined in Figure 2.13 in an ideally plastic material solved using triangular elements

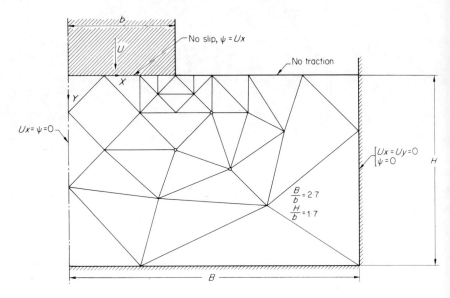

Figure 2.13 Non-steady state problem of punch indentation (40 elements, 29 nodes) ideally plastic material $\bar{\mu} = 0$

and a stream function formulation. In Figure 2.15 the variation of indenting forces with depth of penetration is indicated. Again here the solution is independent of velocity of indentation as $\bar{\mu} = 0$ was assumed. For true viscoplastic materials the solution could be obtained equally easily and would show (as in the previous example) a velocity dependence.

The process of adjusting the free surface profile in successive time steps resembles that used previously for transient seepage solutions.[24,25] Improvements in that context have recently been made by the introduction of a predictor-corrector process in free surface extrapolation and they permit larger time steps to be taken without instability.

A note should be made here of the necessity of periodic mesh shape adjustments as the process continues. These are necessitated by drastic surface configuration changes which are clearly indicated on Figure 2.14.

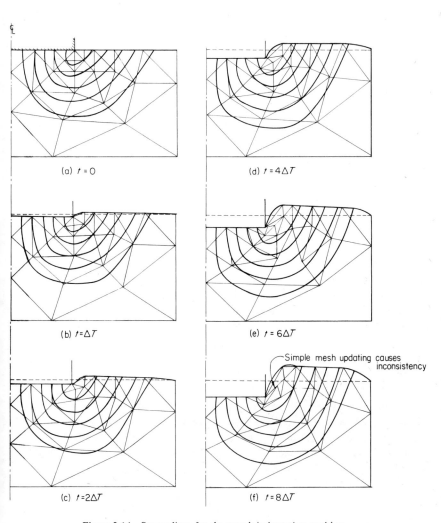

(a) $t = 0$

(b) $t = \Delta T$

(c) $t = 2\Delta T$

(d) $t = 4\Delta T$

(e) $t = 6\Delta T$

Simple mesh updating causes inconsistency

(f) $t = 8\Delta T$

Figure 2.14 Streamlines for the punch indentation problem

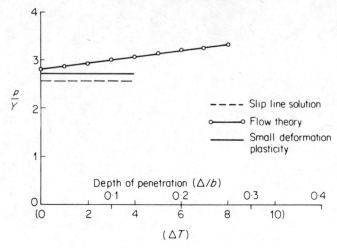

Figure 2.15   Indenting pressure versus penetration depth, Δ

2.10   Inclusion of dynamic effects: Navier–Stokes problem

It was shown in Section 2.3.2 how the inclusion of the dynamic term in the virtual work discretization gave rise to a non-linearity and to non-symmetric matrices. The lack of symmetry is due to the absence of a variational principle for full viscous flow problems.

The problem presented is known to have unique solution at low Reynolds numbers but for large Reynolds number steady state solutions apparently do not exist and the flow becomes turbulent—a fact found experimentally. Solution to the steady state problem will thus only be sought for fairly low velocity flows. With the transient term $\partial u/\partial t$ (see Equation 2.5) included in principle, solution could be obtained for any velocity but, due to the nature of turbulence and its rapid velocity changes, numerical errors could well be expected unless both time and space subdivisions are very fine.

In the low velocity steady state problem we can proceed numerically in two ways: either by isolating the non-linearity in the 'force' term and writing equations corresponding to (2.24) as

$$\bar{\mathbf{K}}_0 \mathbf{a} + \bar{\mathbf{P}} = 0 \tag{2.62}$$

in which $\mathbf{K}_0$ is the symmetric, constant, matrix and $\bar{\mathbf{P}}$ is dependent on velocity and hence

$$\bar{\mathbf{P}} = \bar{\mathbf{P}}(\mathbf{a}) \tag{2.63}$$

An iteration of the type

$$\mathbf{a}^n = \mathbf{K}_0^{-1} \bar{\mathbf{P}}(\mathbf{a}_{n-1}) \tag{2.64}$$

is effective, however, only at very low velocities and at higher ones does not converge. Alternatively we use the non-linear form

$$\mathbf{Ka} + \bar{\mathbf{P}}_0 = 0 \qquad (2.65)$$

in which

$$\mathbf{K} = \mathbf{K(a)} \qquad (2.66)$$

and iterate as

$$\mathbf{a}^n = \mathbf{K(a}_{n-1})\bar{\mathbf{P}}_0 \qquad (2.67)$$

This process has proved effective for quite high velocities (Reynolds numbers) but is obviously more costly requiring repeated inversion of a non-symmetric matrix which has to be recalculated at each stage.

Formulations used so far in this problem include both the approaches outlined in Sections 2.3.2 and 2.3.3, i.e. the use of velocities and pressures as unknown or the use of stream function (in two-dimensional situations).

Reference 3 shows some problems solved using the first type of formulation and an interpolation of C^0 continuity using isoparametric, parabolic elements.

In non-Newtonian situations it is likely that only relatively small effects of inertia will be present. For such situations a combination of iteration of the type given by Equations 2.62 to 2.64 combined with incorporation of changes in K_0, as in Equations 2.47 to 2.49 proves effective and indeed can be accomplished at little more cost than either non-linearity treated independently. Examples of such application will be published elsewhere.

2.11 Concluding remarks

In this chapter we have indicated very many alternatives possible for the solution of viscous flow problems with Newtonian or non-Newtonian viscosity. Some of the possibilities have not yet been explored and many others have not been mentioned. We hope, however, that sufficient indication of the possible applications and relative merits of various approaches has been given.

From the physical viewpoint we have limited the discussion to several relatively simple situations. Possibilities, and indeed the need, for extending the solution technique to new situations are being explored. Here we mention but a few:

(a) Coupled thermal/flow problems where viscosity is temperature dependent.
(b) Problems where surface tension effects are present.
(c) Problems in which straining history causes anisotropy of flow and changes in properties.

References

1. O. C. Zienkiewicz, *The Finite Element Method in Engineering Science*, McGraw-Hill, London, 1971.
2. L. R. Herrmann, 'Elasticity equations for incompressible or nearly incompressible materials by a variational theorem', *AIAA J.*, **3**, 1896 (1965).
3. C. Taylor and P. Hood, 'A numerical solution of the Navier–Stokes equations using the finite element technique', *Computer and Fluids*, **1**, 73–100 (1973).
4. J. T. Oden (ed.), 'The finite element method in fluid mechanics', lecture for NATO Advanced Institute on finite element method in *Continuum Mechanics*, Lisbon, Sept. 1971, University of Alabama Press, 1973.
5. O. C. Zienkiewicz and P. N. Godbole, 'Incompressible elastic materials. A stream function approach to FEM solution' (to be published).
6. O. C. Zienkiewicz and P. N. Godbole, 'Flow of plastic and visco-plastic solids with special reference to extrusion and forming processes', *Int. J. N. Meth. Eng.*, **8**, 3–16 (1974).
7. B. Fraeijs de Veubeke and O. C. Zienkiewicz, 'Strain energy bounds in finite elements by slab analogy', *J. Strain Analysis*, **2**, 265–271 (1967).
8. G. Sander, 'Application of the dual analysis principle', *Proc. IUTAM Symposium on high speed computing of elastic structures, University of Liege, 1970*, pp. 167–209.
9. B. Taborrock, 'A variational principle for the dynamic analysis of continua by hybrid finite element method', *Int. J. Solids Struct.*, **7**, 251–268 (1971).
10. O. C. Zienkiewicz, 'Constrained variational principles and penalty function methods in finite element analysis', *Conf. on Numerical Solution of Differential Equations, Dundee, 1973* (to be published, Springer).
11. B. Atkinson, M. P. Brocklebank, C. C. M. Card and J. M. Smith, 'Low Reynolds number developing flows', *A.I.Ch.E.J.*, **15**, 548–563 (1969).
12. B. Atkinson, C. C. M. Card and B. M. Irons, 'Application of the finite element method to creeping flow problems', *Trans. Inst. Ch. Eng.*, **48**, 276–284 (1970).
13. F. K. Bogner, R. L. Fox and L. A. Schmidt, 'The generation of interelement compatible stiffness and mass matrices by the use of interpolation formulae', *Proc. Conf. Matrix Meths. Structural Mech., Wright-Patterson AFB, Ohio* (1968).
14. H. S. Lew and Y. C. Fung, 'On low Reynolds number entry flow into a circular tube', *Journal of Bio-Mechanics*, **2**, 105–119 (1969).
15. L. S. Han, 'Hydrodynamic entrance lengths for incompressible flow in rectangular ducts', *J. of Applied Mechanics Trans. A.S.M.E.*, **82**, Series E, 403–409 (1960).
16. G. A. Carlson and R. W. Hornbeck, 'A numerical solution for laminar entrance flow in a square duct', *J. of Applied Mechanics T.A.S.M.E.*, Series E, 25–30 (1973).
17. G. P. Bazeley, Y. K. Cheung, B. M. Irons and O. C. Zienkiewicz, 'Triangular elements in bending—conforming and non-conforming solutions', *Proc. Conf. Matrix Methods in Struct. Mechanics, Air Force Inst. of Tech. Wright-Patterson AFB, Ohio* (1965).
18. A. Razzaque and B. M. Irons, 'Shape function formulation for elements other than displacement models', *Int. Conf. on Variational Methods in Eng., Southampton*, September 1972.
19. A. Razzaque, 'Program for triangular bending elements with derivative smoothing', *Int. J. for Num. Methods in Eng.*, **6**, 333–343 (1972).
20. K. Palit and R. T. Fenner, 'Finite element analysis of two dimensional slow non-Newtonian flows', *A.I.Ch.E. Journal*, **18**, 1163–1169 (1972).
21. K. Palit and R. T. Fenner, 'Finite element analysis of slow non-Newtonian channel flow', *A.I.Ch.E. Journal*, **18**, 628–633 (1972).

22. P. Perzyna, 'Fundamental problems in visco-plasticity', *Recent Advances in Applied Mechanics*, Academic Press, New York, Vol. 9, 1966, pp. 243–377.
23. O. C. Zienkiewicz and I. C. Cormeau, 'Visco-plasticity solution by finite element process', *Archives Mechanics* (*Poland*), **24**, 873–889 (1972).
24. C. Taylor, P. W. France and O. C. Zienkiewicz, 'Some free surface transient flow problems of seepage and irrotational flow', in *The Mathematics of Finite Elements and Applications*, ed. T. R. Whiteman, Academic Press, 1973, pp. 313–325.
25. P. W. France, C. J. Parekh, T. C. Peters and C. Taylor, 'Numerical analysis of free surface seepage problems', *Proc. A.S.C.E. I.R.1*, 165–179 (1971).

Chapter 3

Variational-Finite Element Methods for Two-Dimensional and Axisymmetric Navier–Stokes Equations

M. D. Olson

3.1 Introduction

The work of applying the finite element method to fluid mechanics has progressed through the simplest linear inviscid type problems[1,2] to slow viscous flow problems[3,4,5] and finally to the solution of the full Navier–Stokes equations.[6,7,8,9] However, this latter area represents an extremely large and complex field. As such the potential work here has only just begun.

In the present work pseudo-variational finite element theories are formulated for steady incompressible flow in both two-dimensional and axisymmetric geometries. The approach uses the stream function ψ as the dependent variable and incorporates a high precision 18 degree of freedom triangular finite element which provides continuity of ψ and its first derivatives. This guarantees convergence as more and more elements are used. Special care is taken to include boundary conditions in the formulation to fully utilize the potential of the method. The resulting non-linear algebraic equations are solved by a Newton–Raphson method. The formulations are tested on several examples and typical results are presented.

3.2 Theoretical formulation

3.2.1 Conservation equations

Steady, incompressible flow in two-dimensional or axisymmetric geometry is considered in the present work. The basic equations for these problems are the well-known Navier–Stokes equations[10] which are written using a suffix notation to denote differentiation, as:

$$p_{,x} + \rho(uu_{,x} + vu_{,y}) = \mu(u_{,xx} + u_{,yy}) \tag{3.1a}$$

$$p_{,y} + \rho(uv_{,x} + vv_{,y}) = \mu(v_{,xx} + v_{,yy}) \tag{3.1b}$$

$$u_{,x} + v_{,y} = 0 \tag{3.1c}$$

for two-dimensional flow and

$$p_{,x} + \rho(uu_{,x} + vu_{,r}) = \mu(u_{,xx} + u_{,rr} + u_{,r}/r) \tag{3.2a}$$

$$p_{,r} + \rho(uv_{,x} + vv_{,r}) = \mu(v_{,xx} + v_{,rr} + v_{,r}/r - v/r^2) \tag{3.2b}$$

$$(ru)_{,x} + (rv)_{,r} = 0 \tag{3.2c}$$

for axisymmetric flow, where u, v are the x, y or x, r components of velocity, ρ is the fluid density, p the pressure, and μ the absolute viscosity. Equations labelled (a) and (b) represent conservation of momentum and equations labelled (c) represent conservation of mass in the two geometries. Since Newtonian fluid has been assumed, the shear stress is given by

$$\tau_{xy} = \mu(u_{,y} + v_{,x}) \tag{3.3a}$$

and

$$\tau_{xr} = \mu(u_{,r} + v_{,x}) \tag{3.3b}$$

for the two geometries. The classical reduction of Equations 3.1 and 3.2 is obtained by introducing stream functions ψ defined by

$$u = -\psi_{,y}, \qquad v = \psi_{,x} \tag{3.4a}$$

and

$$ru = -\psi_{,r}, \qquad rv = \psi_{,x} \tag{3.4b}$$

for the two cases, respectively, thereby satisfying the continuity equations (3.1c) and (3.2c) exactly. Elimination of the pressure p from each set then leads for constant viscosity to single equations for the stream function ψ, namely

$$\nu\nabla^4\psi + \psi_{,x}\zeta_{,y} - \psi_{,y}\zeta_{,x} = 0 \tag{3.5a}$$

for the two-dimensional case and

$$\frac{\nu}{r}[\psi_{,xxxx} + 2\psi_{,xxrr} + \psi_{,rrrr} - 2\psi_{,rrr}/r + 3\psi_{,rr}/r^2 - 3\psi_{,r}/r^3 - 2\psi_{,xxr}/r]$$

$$+ \frac{1}{r}[\psi_{,x}(\zeta_{,r} - \zeta/r) - \psi_{,r}\zeta_{,x}] = 0 \tag{3.5b}$$

for the axisymmetric one, where ζ is the vorticity

$$\zeta = \nabla^2\psi = \psi_{,xx} + \psi_{,yy} \tag{3.6a}$$

in the former case, and

$$\zeta = (\psi_{,xx} + \psi_{,rr} - \psi_{,r}/r)/r \tag{3.6b}$$

in the latter. v is the kinematic viscosity μ/ρ. The non-linear convection terms in these equations make them very difficult to solve exactly but approximate solutions may be obtained once each problem is discretized.

3.2.2 'Variational' principles

The non-linear terms in the Navier–Stokes equations seem to have precluded the existence of an associated variational principle of the classical kind. However, it has been shown[11] that ad hoc or 'pseudo' principles can be obtained provided some terms are not allowed to vary when the first variation is performed. Whilst this leads to a similar formulation as that obtained using a Galerkin principle (see Chapters 2, 11 and 14) it is a convenient alternative for present purposes.

Multiplying Equations (3.5) by 'virtual variations' $\delta\psi$ and integrating over the areas $dx\,dy$ and $dx\,dr$ respectively, yields

$$\delta\Pi_p = \iint [v\nabla^4\psi + \psi_{,x}\zeta_{,y} - \psi_{,y}\zeta_{,x}]\,\delta\psi\,dx\,dy \tag{3.7a}$$

and

$$\delta\Pi_a = \iint \left\{ \frac{v}{r}[\psi_{,xxxx} + 2\psi_{,xxrr} + \psi_{,rrrr} - 2\psi_{,rrr}/r + 3\psi_{,rr}/r^2 \right.$$
$$\left. - 3\psi_{,r}/r^3 - 2\psi_{,xxr}/r] + \frac{1}{r}[\psi_{,x}(\zeta_{,r} - \zeta/r) - \psi_{,r}\zeta_{,x}]\right\}\,\delta\psi\,dx\,dr \tag{3.7b}$$

for plane and axisymmetric situations respectively.

The next step is to integrate these equations by parts sufficient times to produce a 'symmetric' functional with minimum order of derivatives. The result is far from unique but many possibilities may be eliminated according to various constraints, the most obvious being that logical boundary conditions should be generated when the first variation is set to zero. The final results achieved herein are

$$\Pi_p = \iint \left[\frac{v}{2}(\nabla^2\psi)^2 + \underline{(\psi_{,y}\zeta)\psi_{,x}} - \underline{(\psi_{,x}\zeta)\psi_{,y}}\right] dx\,dy \tag{3.8a}$$

and

$$\Pi_a = \iint \left[\frac{v}{2r}(\psi_{,xx} + \psi_{,rr} - \psi_{,r}/r)^2 + \underline{(\psi_{,r}\zeta/r)\psi_{,x}} - \underline{(\psi_{,x}\zeta/r)\psi_{,r}}\right] dx\,dr \tag{3.8b}$$

Π_p and Π_a are then functionals which will yield the differential Equations 3.5a and 3.5b respectively, when their first variations are set equal to zero, *provided* the underlined bracketed terms are held constant during the variations. The associated boundary integrals which are generated simultaneously are

$$\nu \int [\zeta(\nabla\,\delta\psi).\mathbf{n} - \nabla\zeta.\mathbf{n}\,\delta\psi] + \int (\nabla\psi.\mathbf{s})\zeta\,\delta\psi\,ds \qquad (3.9a)$$

where $\mathbf{n}$ is a unit outward normal to the boundary and $\mathbf{s}$ is the unit tangent vector, and

$$\int [\nu\zeta\,\delta\psi_{,x} - (\nu\zeta_{,x} - \zeta\psi_{,r}/r)\,\delta\psi]\,dr$$

$$+ \int [\nu\zeta\,\delta\psi_{,r} - (\nu\zeta/r + \nu\zeta_{,r} + \zeta\psi_{,x}/r)\,\delta\psi]\,dx \qquad (3.9b)$$

respectively. These boundary integrals will then vanish for the following boundary conditions in two dimensions: (i) either $\delta\psi = 0$ or $\zeta\nabla\psi.\mathbf{s} - \nu\nabla\zeta.\mathbf{n} = 0$ and (ii) either $\nabla\,\delta\psi.\mathbf{n} = 0$ or $\zeta = 0$. The boundary conditions for the axisymmetric case are completely analogous, although somewhat less neat. The 'rigid' boundary conditions of $\delta\psi = 0$ and $\nabla\,\delta\psi.\mathbf{n} = \delta(\partial\psi/\partial n) = 0$ are completely logical, but the alternative 'natural' boundary conditions are somewhat arbitrary, since apparently there is little agreement in the field of fluid mechanics as to what the 'proper' ones are. The choice therefore is made on the basis of practicality. For instance, far from any disturbance in a constrained channel or pipe flow the non-linear terms are negligible and therefore the first natural boundary condition reduces to $\nabla\zeta.\mathbf{n} = 0$, which is equivalent to $\partial p/\partial s = 0$ (where s is tangential to the boundary). The second natural condition of zero vorticity needs no explanation.

In the work previously reported,[8] the foregoing conditions were actually converted rigorously to $\partial p/\partial s = 0$ and $\tau_{xy} = 0$ respectively. This was done by adding an appropriate boundary integral along one edge of the triangular finite element to the variational principle. This edge was then used along all boundaries where the natural boundary conditions were to be satisfied.

In the present axisymmetric theory, the natural boundary condition of $\partial p/\partial s = 0$ is included by adding the following integral to the right side of Equation 3.8b.

$$\Pi_a = \frac{1}{2}\iint \{[(\psi_{,x}/r)^2 + (\psi_{,r}/r)^2]_{,x}\psi_{,r} - [(\psi_{,x}/r)^2 + (\psi_{,r}/r)^2]_{,r}\psi_{,x}\}\,dx\,dr \qquad (3.10)$$

where again the underlined terms are not varied. Both versions of Π_a have been used to solve certain numerical examples.

It is observed that the first term of Π_p and Π_a in Equations 3.8 which represent linear flow are simply spatial integrals of the vorticity squared. That is,

$$\frac{\nu}{2}\iiint_0^1 \zeta^2\,dx\,dy\,dz \qquad (3.11a)$$

and

$$\frac{v}{4\pi} \iiint_0^{2\pi} \zeta^2 r \, dx \, dr \, d\theta \qquad (3.11b)$$

Seeking the extreme values of these integrals leads to the correct linear differential equations and the boundary conditions that (i) either $\delta\psi = 0$ or $\partial p/\partial s = 0$ and (ii) either $\delta\psi_{,n} = 0$ or $\zeta = 0$. It may be noted that the functional of equations 3.11 appears different from the 'classical' one used by Atkinson and coworkers[12] and obtained by minimizing the dissipation.

3.2.3 Finite element formulations

Both functionals Π_p and Π_a contain derivatives of ψ up to the second order and therefore continuity of ψ and its first derivatives between elements is needed to guarantee convergence to the correct solution as the number of elements is increased C^1 continuity. The high precision triangular element[13] satisfying this requirement is used to represent ψ in both geometries. The element uses ψ, its two first derivatives and its three second derivatives as generalized coordinates at each vertex yielding a total of 18 degrees of freedom. The formulation begins with a full fifth degree polynomial with 21 parameters and three constraints are imposed to ensure that the slope normal to an edge ($\partial\psi/\partial n$) varies only as a cubic along that edge. This leaves 18 parameters with the 18 corner degrees of freedom and ensures continuity of ψ and $\partial\psi/\partial n$ between adjacent elements provided appropriate degrees of freedom are made equal at the vertices.

The derivation for the two-dimensional case is worked out most easily in the local coordinates ξ, η shown in Figure 3.1. The detail which follows standard finite element methods may be found in Reference 8. On the other

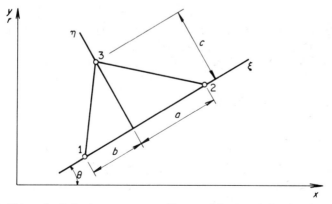

Figure 3.1 Triangular finite element geometry. Degrees of freedom ψ, $\psi_{,x}$, $\psi_{,y}$, $\psi_{,xx}$, $\psi_{,yy}$, $\psi_{,xy}$ at each node

hand, the axisymmetric case is handled in the global x, r system. That is, ψ is taken as the full quintic polynomial

$$\psi = \sum_{i=1}^{21} a_i x^{m_i} r^{n_i} \qquad (3.12)$$

where the vectors m_i, n_i contain the appropriate exponents. The polynomial coefficient a_i are then transformed to the vertex degrees of freedom by writing

$$\{\Psi, 0, 0, 0\}^{\mathrm{T}} = \mathbf{Ta} \qquad (3.13)$$

where

$$\boldsymbol{\psi} = (\psi_1, \psi_{x1}, \psi_{r1}, \psi_{xx1}, \psi_{xr1}, \psi_{rr1}, \psi_2, \ldots, \psi_{rr3}) \qquad (3.14)$$

with letter subscripts denoting derivatives and number subscripts denoting vertices 1 to 3. $\mathbf{a}$ is the vector of polynomial coefficient a_1 to a_{21} and $\mathbf{T}$ is the transformation relating them to Ψ. The first 18 rows are simply evaluations of the 6 degrees of freedom at each of the 3 triangle vertices. The last 3 rows contain the constraint equations

$$5\cos^4 \beta \sin \beta a_{16} + (4\cos^3 \beta \sin^2 \beta - \cos^5 \beta)a_{17}$$

$$+ (3\cos^2 \beta \sin \beta - 2\cos^4 \beta \sin \beta)a_{18}$$

$$+ (2\sin^4 \beta \cos \beta - 3\cos^3 \beta \sin^2 \beta)a_{19}$$

$$+ (\sin^5 \beta - 4\cos^2 \beta \sin^3 \beta)a_{20}$$

$$- 5\cos \beta \sin^4 \beta a_{21} = 0 \qquad (3.15)$$

where β is the angle between the side and the x axis. Equation 3.13 is inverted to yield

$$\mathbf{a}^{\mathrm{T}} = \mathbf{T}^{-1}\{\Psi, 0, 0, 0\} = \mathbf{S}\boldsymbol{\Psi} \qquad (3.16)$$

where $\mathbf{S}$ contains the first 18 columns of $\mathbf{T}^{-1}$. The element properties are obtained first with respect to the polynomial coefficients by substituting Equation 3.12 into Equation 3.8b and integrating. This gives

$$\Pi_a = \frac{\nu}{2}\sum_i^{21}\sum_j^{21} k_{ij}a_i a_j + \sum_i^{21}\sum_j^{21}\sum_k^{21} g_{ijk}\bar{a}_i\bar{a}_j a_k \qquad (3.17)$$

where

$$k_{ij} = m_i m_j(m_i - 1)(m_j - 1)\mathscr{F}(m_i + m_j - 4, n_i + n_j - 1)$$

$$+ [m_i n_j(m_i - 1)(n_j - 2) + m_j n_i(m_j - 1)(n_j - 2)]$$

$$\times \mathscr{F}(m_i + m_j - 2, n_i + n_j - 3)$$

$$+ n_i n_j(n_i - 2)(n_j - 2)\mathscr{F}(m_i + m_j, n_i + n_j - 5) \qquad (3.18)$$

and

$$q_{ijk} = (n_i m_k - m_i n_k)[m_j(m_j - 1)\mathscr{F}(m_i + m_j + m_k - 3, n_i + n_j + n_k - 3)$$
$$+ n_j(n_j - 2)\mathscr{F}(m_i + m_j + m_k - 1, n_i + n_j + n_k - 5)] \qquad (3.19)$$

The bars over the a_i in Equation 3.17 mean that those terms are not varied when the first variation is taken. The function $\mathscr{F}(i, j)$ is defined by

$$\mathscr{F}(i, j) = \iint x^i r^j \, dx \, dr \qquad (3.20)$$

where the integration is over the area of the triangular element. This integral is carried out exactly as the sum over three trapezoidal areas with the x axis as the bottom side, two sides parallel to the r axis and the fourth side as one side of the triangle. The analytical result was programmed as a subroutine.

The element properties are then transformed to global coordinates by combining Equations 3.16 and 3.17 to give

$$\Pi_a = \frac{v}{2}\sum_i^{18}\sum_j^{18} k_{ij}\Psi_i\Psi_j + \sum_i^{18}\sum_j^{18}\sum_k^{18} G_{ijk}\overline{\Psi}_i\overline{\Psi}_j\Psi_k \qquad (3.21)$$

where

$$K_{ij} = \sum_r^{21}\sum_s^{21} S_{ri}S_{sj}k_{rs} \qquad (3.22)$$

and

$$G_{ijk} = \sum_r^{21}\sum_s^{21}\sum_t^{21} S_{ri}S_{sj}S_{tk}g_{rst} \qquad (3.23)$$

may be thought of as the linear and non-linear stiffness matrices respectively, for the finite element. Note that **K** is completely symmetric while **G** is symmetric only in i and j. Again, the bars over the Ψ_i mean: not varied during the variation.

The foregoing axisymmetric formulation requires some further consideration because the region near $r = 0$ has special characteristics. The pertinent physical conditions to be satisfied there are $\psi_{,r} = 0$ and $\psi_{,r} - r\psi_{,rr}$ be of order r^2. Satisfaction of these conditions ensures the integrability of Equation 3.8b throughout the domain. Unfortunately, the triangular finite element used herein does not satisfy these conditions for an element with only one node on the x-axis although it does for an element with two nodes there. (These facts were determined from a numerical investigation of the inverted transformation matrix in Equation 3.16. This difficulty could possibly be overcome by the introduction of a special element on the axis satisfying these conditions. However an alternative approach was taken here. That is, all singular integrations at $r = 0$ were arbitrarily suppressed in the sub-routine used to evaluate the $\mathscr{F}(i, j)$ of Equation 3.20. This somewhat crude approach appears to be satisfactory in elasticity problems[14] and it was hoped that it would be so here as well. It has the effect of destroying the completeness of the polynomial distributions in all elements along the axis, but any accompanying error should vanish as the element size goes to zero.

3.2.4   Assemblage and solution of non-linear equations

In solving a fluid flow problem with the foregoing elements, the usual assemblage process for finite elements[14] is followed here as well. That is, a finite element gridwork is established for the given problem and the individual stiffness matrices from each element are appropriately summed into the global matrices for the complete problem. The non-linear matrix **G** of Equation 3.23 is handled just like the linear one **K** is, the only difference being that the summations extend over an extra dimension. The homogeneous boundary conditions are introduced during this process, so that only non-zero terms are retained in the final equations. This results in a global representation of the functional Π_p or Π_a for the complete problem, i.e.

$$\Pi_p = \frac{\nu}{2} \sum_i^n \sum_j^n K_{ij} \Psi_i \Psi_j + \sum_i^n \sum_j^n \sum_k^n G_{ijk} \overline{\Psi}_i \overline{\Psi}_j \Psi_k \qquad (3.24)$$

where for simplicity of notation **K** and **G** now represent the global matrices of size n. (The non-linear matrix **G**, because of its 'three-dimensional size', must be stored out of core, the high-speed disc being the next best place. This was done in the present work taking full account of symmetry and bandedness. For instance, it was required to store a rectangular array of NB by $2NB - 1$ for the general kth slice of G_{ijk}, where NB was the half-band width for the problem.)

The solution of any problem is defined by seeking the extremum of Equation (3.24), while holding the barred variables fixed. This results in

$$\nu \sum_j^n K_{kj} \Psi_j + \sum_i^n \sum_j^n G_{ijk} \Psi_i \Psi_j = 0 \qquad (3.25)$$

for $k = 1$ to n, and from now on there is no difference between barred and unbarred variables. Equations 3.25 then form a set of n non-linear algebraic equations for the n global variables Ψ_i. These equations may be solved by the Newton–Raphson method as follows. Denote the kth equation in Equation 3.25 by R_k and let Ψ_{i0} be an approximation of the correct solution Ψ_i. Then by Taylor's series

$$R_k(\Psi_{i0}) + \sum_{l=1}^n [\partial R_k / \partial \Psi_l] \Delta \Psi_l \simeq 0 \qquad (3.26)$$

where the gradient of R_k is

$$\partial R_k/\partial \Psi_i = \nu K_{ki} + \sum_{j=1}^{n} (G_{ijk} + G_{jik})\Psi_{j0} \qquad (3.27)$$

and $\Delta \Psi_i$ is the difference between solutions, i.e.

$$\Psi_i = \Psi_{i0} + \Delta \Psi_i \qquad (i = 1, n) \qquad (3.28)$$

$\Delta \Psi$ is determined by solving the linear equations 3.26 and 3.28, gives the next approximation. The whole process is repeated until any desired accuracy is reached.

It may be noted that Equations 12–25 are homogeneous. This is in contrast to most static problems in structural mechanics where there would be load terms on the right-hand side. In the present context, the 'loading' comes from some of the boundary conditions. For example, in the case of a uniform flow upstream, the stream function ψ and its derivatives are known and this information is introduced as a constraint. This is easily done by specifying all the non-zero known nodal variables Ψ_i as constrained and introducing their exact values into the first solution guess. Before solving Equation 3.26, the rows and right-hand sides corresponding to the constrained variables are zeroed and 1s are placed on the diagonal. This provides the correct answers for all $\Delta \Psi_i$ including $\Delta \Psi_i = 0$ for all constrained variables.

3.2.5 Stream lines and pressure field

Once all the nodal variables have been found by the foregoing process, further interpolative results can readily be obtained within each element. The first step is to calculate the polynomial coefficients from the nodal variables for the particular element of interest from Equation 3.16. Then various quantities can be obtained from the stream function polynomial. For instance, stream function and velocity levels may be calculated anywhere within the element and used for obtaining contours.

In Reference 8, the pressure field was obtained by integrating the momentum equations using the above known stream function. It may alternatively be obtained by solving the Poisson type equations

$$\frac{\nabla^2 p}{\rho} = 2(\psi_{xx}\psi_{yy} - \psi_{xy}^2) \qquad (3.29a)$$

in two dimensions, and

$$\frac{\nabla p}{\rho} = (\psi_{,x}\zeta)_{,x}/r + (\psi_{,r}\zeta)_{,r}/r - \tfrac{1}{2}\nabla^2[(\psi_{,x}/r)^2 + (\psi_{,r}/r)^2] \qquad (3.29b)$$

in axisymmetric flow. These equations are readily solvable by the finite element method.[14] For instance, denoting the right-hand side of Equation

3.29b by $\mathcal{Q}(x, r)$, the equation may be replaced by its equivalent variational integral

$$\Pi_c = \tfrac{1}{2} \iint [(p_{,x}^2 + p_{,r}^2) + \rho \mathcal{Q}(x, r)p]r \, \mathrm{d}x \, \mathrm{d}r + \int r(\nabla p \cdot \mathbf{n})p \, \mathrm{d}s \qquad (3.30)$$

Minimizing Π_c then is a standard linear finite element problem with a given distributed load $\mathcal{Q}(x, r)$ over the domain and given normal pressure gradient on the boundary wherever it is specified.

3.3   Example applications

The above pseudo-variational-finite element formulations have been tested on several numerical examples. Some of the results are shown and summarized in the following. More extensive two-dimensional results may be found in Reference 8.

3.3.1   Fully developed parallel flow

The first test case was that of fully developed flow in a two-dimensional channel and in an axisymmetric tube. The finite element grids used are shown in Figure 3.2. These problems have the exact solutions $\psi = 3y^2 - 2y^3$ and

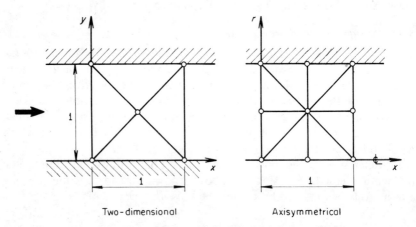

Two-dimensional   Axisymmetrical

Figure 3.2   Finite element grids for fully developed parallel flow

$\psi = r^2(1 - r^2/2)$ respectively. The boundary conditions imposed were as follows: for two dimensions, (a) $\psi = 3y^2 - 2y^3$, $\psi_{,x} = 0$ on the upstream edge, (b) $\psi = 0$, $\psi_{,y} = 0$ on the bottom edge, (c) $\psi = 1$, $\psi_{,y} = 0$ on the top edge and (d) $\psi_{,x} = 0$ on the downstream edge; and for axisymmetric flow, (a) $\psi = r^2(1 - r^2/2)$, $\psi_{,x} = 0$ on the upstream edge, (b) $\psi = 0$, $\psi_{,r} = 0$ along the centre-line axis, (c) $\psi = 0.5$, $\psi_{,r} = 0$ on the top edge and (d) $\psi_{,x} = 0$ on the downstream edge. Note that all these conditions are 'rigid' ones in the sense of Section 3.2.2. The program then was left to seek its own approximation of the natural boundary condition $p_{,y} = p_{,r} = 0$ (zero pressure gradient across the flow) on the downstream edge.

The ensuing calculations, starting with all free variables nulled, reproduced the exact solution on the downstream edge in a few iterations, the number of iterations required increasing slightly with Reynolds number. The result that the downstream natural boundary conditions were represented exactly here was to be expected since the present elements contain complete quartic polynomials for ψ and hence can represent the correct solution exactly. Normally, many elements would be required to reasonably approximate such natural boundary conditions.

3.3.2 Flow over a circular cylinder

The next example considered was the classical problem of two-dimensional flow over a circular cylinder crosswise to the flow. This problem was of special interest here because, above a certain critical Reynolds number, the flow separates from the downstream side of the cylinder and forms a recirculating vortex. Prediction of this separation provided a telling test for the present method. Further, good finite difference results by Dennis and Chang[15] were available here as a guide.

The physical situation is that of an infinitely long circular cylinder of unit diameter immersed in a fluid medium of infinite extent. The flow far

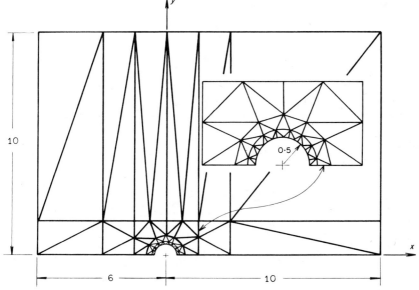

Figure 3.3 Finite element grid for flow over a circular cylinder, $NF = 202$, $NB = 57$

upstream is uniform. The finest element grid used to represent this is shown in Figure 3.3. The rigid boundary conditions imposed exactly were as follows: (a) $\psi = y$, $\psi_{,x} = 0$ on the upstream edge, (b) $\psi = 0$ on the x axis, (c) $\psi_{,y} = 1$ along the upper edge and (d) $\psi_{,x} = 0$ on the downstream edge. The rigid boundary conditions of $\psi = 0$ and $\psi_{,n} = 0$ (n being normal to the cylinder) could not be satisfied exactly and hence were approximated by the method given in Reference 16. That is, for nodes on the cylinder surface, the nodal variables were transformed from the global x, y coordinate system to a local t, n (tangential, normal) system and all variables except $\psi_{,nn}$ were zeroed and eliminated. This had the effect of providing the desired boundary conditions exactly only at discrete points on the cylinder, namely at the nodes. However, the results of Reference 16 indicated that a reasonable approximation of the same conditions would also be provided between the nodes. Finally, the natural boundary conditions left for the program to approximate were then zero shear stress along the symmetry line (bottom edge) and zero pressure gradient along both the top and downstream edges.

The Reynolds number based on the cylinder diameter was then $1/\nu$. The calculations were carried out for $Re = 0.1$, 1, 5, 7, 10, 20, 40 and 70 and 100 and never more than four or five iterations were required. The streamline patterns obtained for $Re = 20$ are shown in Figure 3.4 along with the finite difference ones of Dennis and Chang.[15] The separated flow patterns compare very well, especially as to the point of separation. The stream function values

Finite elements

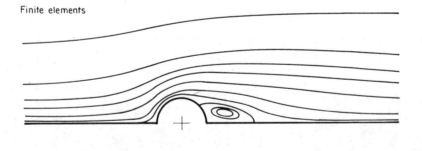

Finite differences

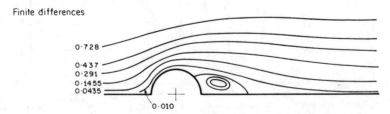

Figure 3.4 Streamlines over circular cylinder, $Re = 20$

therein are shown bracketed. The results for other Reynolds numbers were also confirmed. However, the numerical results for quantities on the cylinder surface such as drag and pressure coefficient did not compare so well. This was due to the boundary condition approximations there and could best be cured with a curved edge finite element. It may be noted that Dennis and Chang used 3000 equations compared to 202 here.

The present program used about 2 minutes of IBM 360/67 CPU time per iteration for this example.

3.3.3 *Flow in a constricted cylindrical tube*

The next example considered was of axisymmetric flow in a tube with a step change in diameter. The finite difference results of Greenspan[17] were used as comparison. The finite element grid used for a section bounded by the wall and the centre-line is shown in Figure 3.5. The rigid boundary

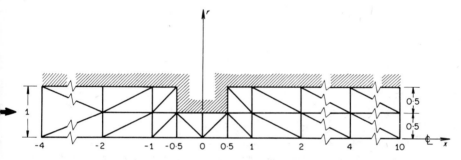

Figure 3.5 Finite element grid for flow in a constricted cylindrical tube

conditions imposed were as follows: (a) $\psi = r^2(2 - r^2)$, $\psi_{,x} = 0$ on the upstream edge (Poiseuille conditions), (b) $\psi = 1$, $\psi_{,r} = 0$ (or $\psi_{,x} = 0$) all along the wall, (c) $\psi = 0$, $\psi_{,r} = 0$ along the centre-line and (d) $\psi_{,x} = 0$ on the downstream edge. The natural condition of $\partial p/\partial r = 0$ was then left for the program to approximate on the downstream edge. The reentrant corners in the step were handled by using extra nodes there with all variables constrained to be equal except for the second derivatives normal to the wall. For instance, this then allowed $\psi_{,xx}$ to be zero for $0.5 < x < 0.5$, $r = 0.5$ but non-zero for $|x| > 0.5$, $r = 0.5$. The $\psi_{,rr}$ was similar.

The Reynolds number based on the larger tube radius was then $1/v$ and the program was run for $Re = 10^{-3}$, 1, 2, 4, 8, 10, 20 and 100. The streamline results for $Re = 10$ are shown in Figure 3.6 along with those of Greenspan[17] using the same contour values. It is seen that the contours compare reasonably well thus indicating that the present axisymmetric program is accurately representing the phenomenon. The slight discrepancies between the two results are probably due to the coarseness of the finite element grid especially

Finite elements

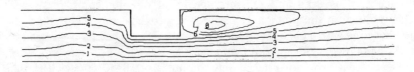

Finite differences

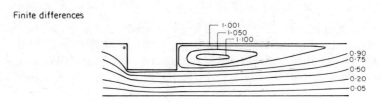

Figure 3.6   Streamlines for flow in a constricted cylindrical tube, $Re = 10$

in the region of the recirculating vortex. Further, the small bubble upstream of the step observed by Greenspan was not obtained here, again because of grid coarseness.

Some of the results for other Reynolds numbers are also given in Figure 3.7. These illustrate the development of the separation vortex which is not present at very low Reynolds numbers. It appears first for Re slightly below 1 and then grows rapidly for Re greater than 1. This was so large for $Re = 100$ that it interfered with the downstream boundary conditions, thus invalidating the result.

3.4   Concluding remarks

The two-dimensional program consistently exhibited good stability, fast convergence and good accuracy, even for the higher Reynolds numbers. However, the axisymmetric one behaved relatively poorly. For example, sometimes it would not converge for higher Reynolds numbers or for more refined grids (in some other examples not reported here). Unfortunately, this behaviour was somewhat erratic so that no consistent pattern has yet emerged.

Acknowledgement

The diligent help of Mr. F. Mirza with the axisymmetric flow work is gratefully acknowledged. This research has been supported by the National Research Council Grant A7730 and Defence Research Board Grant 9550-53.

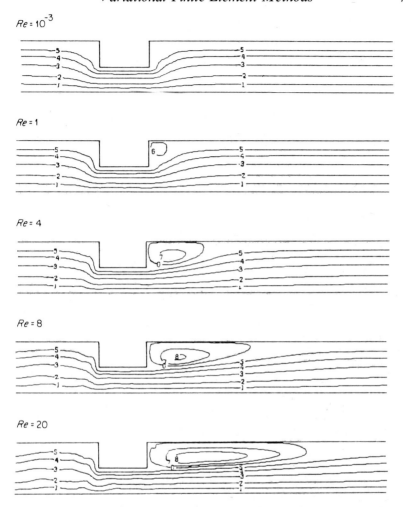

Figure 3.7 Comparison of flow in constricted cylindrical tube for different Reynolds numbers

References

1. H. C. Martin, 'Finite element analysis of fluid flows', *Proc. 2nd Conf. Matrix Methods in Structural Mechanics, Wright-Patterson AFB, Ohio, 1968*, AFFDL-TR-68-150.
2. J. H. Argyris, G. Mareczek and D. W. Scharpf, 'Two and three dimensional flow using finite elements', *Aero J. Roy. Aero. Soc.*, **73**, 961–964 (November 1969).
3. P. Tong, 'The finite element method for fluid flow', Paper US 5-4, *Japan-U.S. Seminar on Matrix Methods of Structural Analysis and Design, Tokyo, Japan, August 1969*.

72 *Finite Elements in Fluids*

4. B. Atkinson, M. P. Brocklebank, C. C. H. Card and J. M. Smith, 'Low Reynolds number developing flow', *A.I. Ch. E. J.*, **15**, 548–553 (1969).
5. J. T. Oden and D. Sornogyi, 'Finite element applications in fluid dynamics', *J. Eng. Mech. Div., Proc. A.S.C.E.*, **95**, No. EM3, 821–826 (1969).
6. J. T. Oden, 'Finite element analogue of Navier–Stokes equations', *J. Eng. Mech. Div., Proc. A.S.C.E.*, **96**, No. EM4, 529–534 (1970).
7. E. Skiba, 'A finite element solution of general fluid dynamics problems—natural convection in rectangular cavities', *M. App. Sci. Civ. Eng. Thesis*, University of Waterloo, Ontario, Canada (April 1970).
8. M. D. Olson, 'A variational finite element method for two-dimensional steady viscous flows', *McGill-EIC Conf. on Fin. Elem. Meth. in Civ. Eng., Montreal, June 1–2, 1972*.
9. J. T. Oden and L. C. Wellford Jr., 'Analysis of flow of viscous fluids by the finite-element method', *AIAA J.*, **10**, 12, 1590–1599 (1972).
10. H. Schlichting, *Boundary Layer Theory*, 4th ed., McGraw-Hill, New York, 1960, p. 58.
11. B. A. Finlayson, *The Method of Weighted Residuals and Variational Principles*, Academic Press, New York, 1972.
12. B. Atkinson, C. C. H. Card and B. M. Irons, 'Application of the finite element method to creeping flow problems', *Trans. Instn. Chem. Engrs.*, **48**, T276–284 (1970).
13. G. R. Cowper, E. Kosko, G. M. Lindberg and M. D. Olson, 'Static and dynamic applications of a high precision triangular plate bending element', *AIAA J.*, **7**, 1957–1965 (1969).
14. O. C. Zienkiewicz, *The Finite Element Method in Engineering Science*, 1st ed., McGraw-Hill, 1967, Sec. 4.2.5. (Also 2nd ed. 1971, Sec. 1.6 and Sec. 3.5.)
15. S. C. R. Dennis and G. Z. Chang, 'Numerical solutions for steady flow past a circular cylinder at Reynolds numbers up to 100', *J. Fluid Mechanics*, **42**, Part 3, 471–489 (1970).
16. M. W. Chernuka, G. R. Cowper, G. M. Lindberg and M. D. Olson, 'Finite element analysis of plates with curved edges', *Int. J. for Num. Meth. in Eng.*, **4**, 49–65 (1972).
17. D. Greenspan, 'Numerical studies of viscous, incompressible flow through an orifice for arbitrary Reynolds number', *Int. J. Num. Meth. in Eng.*, **6**, 489–496 (1973).

Chapter 4

Finite Element Analysis of Steady Fluid and Metal Flow

Y. Yamada, K. Ito, Y. Yokouchi, T. Tamano and T. Ohtsubo

4.1 Introduction

In spite of the great success of the finite element method in structural mechanics, its application to fluid mechanics, particularly to steady flow of viscous fluid, is relatively new and until recently only a few attempts have been made. The situation is well summarized in Olson's paper,[1] itself a breakthrough in the application of the finite element method to the field of fluid mechanics. Olson formulated the procedure in which the stream function is incorporated for the solution of the two-dimensional Navier–Stokes equation of incompressible viscous fluid. Although Olson's paper is characterized by the introduction of a unique functional to be made stationary, the basic formulation can be considered to be parallel to that proposed by Zienkiewicz and coworkers.[2] We note further that the finite element solutions with use of the stream function have been attempted in the work of Atkinson and coworkers.[3]

An alternative solution procedure can be contemplated, i.e. one that solves directly the Navier–Stokes equation which is originally expressed in terms of velocity components u, v, w together with the mean pressure p. This alternative is introduced and applied to the Stokes equation by Zienkiewicz,[4] where the weighted residual method is employed that leads to an unsymmetric element 'stiffness' matrix. A similar procedure was followed by Taylor and Hood[5] who gave it the name 'velocity-pressure formulation' and extended it to the Navier–Stokes equation. By a slight modification, however, stiffness matrices which preserve the symmetric nature can be obtained for the Stokes equation. The present paper elucidates this possibility and emphasizes the versatility of the 'velocity-pressure formulation' by some numerical examples, It should be noted that the present method suggests the existence of a quadratic form of functional to be minimized. Further, our method features the adoption of a quadratic shape function for velocity components and a linear one for pressure distribution. Similar shape functions are used in the paper of Kawahara and coworkers.[6]

The second challenging field of steady flow analysis is that concerned with the continuous processing of metals. The finite element method has not necessarily been successful to date in this field of applications and, in fact, some writers throw doubt on the possibilities of the finite element method and adhere to a preference for the classical slip-line field theory. Significant progress has indeed been achieved in slip-line theory,[7,8,9] but the drawbacks of the theory, which require the intuition and/or skill of the users, have not yet been removed.

The present paper solves the elastic or unsteady elastic-plastic problem by first applying well-established finite element procedures of elasto-plasticity.[10,11] Then, by using the solution as the first approximation and following the change of mechanical state of material elements along assumed respective stream lines, the steady state is pursued iteratively. A solution of strip skin-pass rolling with small thickness reduction is presented as our first successful example. It reveals the characteristic deformation of the rolling process that accompanies intense shear zones in the regions where the material element enters and leaves the roll gap. The previous rolling theories failed to predict these shear zones.

4.2   Analysis of viscous fluid flow

4.2.1   Basic equations and finite element formulation

The Navier–Stokes equation for two-dimensional steady flow of an incompressible fluid with constant coefficient of viscosity μ and density ρ can be written as follows:

$$\left.\begin{array}{l} \mu\left(\dfrac{\partial^2 u}{\partial x^2} + \dfrac{\partial^2 u}{\partial y^2}\right) - \dfrac{\partial p}{\partial x} = \rho\left(u\dfrac{\partial u}{\partial x} + v\dfrac{\partial u}{\partial y}\right) \\[4mm] \mu\left(\dfrac{\partial^2 v}{\partial x^2} + \dfrac{\partial^2 v}{\partial y^2}\right) - \dfrac{\partial p}{\partial y} = \rho\left(u\dfrac{\partial v}{\partial x} + v\dfrac{\partial v}{\partial y}\right) \end{array}\right\} \qquad (4.1)$$

where u and v denote the velocity components and p is the fluid pressure. For simplicity, the body force has been neglected in the above. The incompressibility relation is

$$\frac{\partial u}{\partial x} + \frac{\partial v}{\partial y} = 0 \qquad (4.2)$$

The Stokes equation corresponds to the case where the convective terms on the right-hand side of Equation 4.1 are disregarded. Under this assumption, the governing equations become

$$
\left.
\begin{aligned}
\mu\left(\frac{\partial^2 u}{\partial x^2} + \frac{\partial^2 u}{\partial y^2}\right) - \frac{\partial p}{\partial x} = 0 \\
\mu\left(\frac{\partial^2 v}{\partial x^2} + \frac{\partial^2 v}{\partial y^2}\right) - \frac{\partial p}{\partial y} = 0
\end{aligned}
\right\}
\tag{4.3}
$$

We start our discussion with the Stokes equation and apply the weighted residual method to Equation 4.3 as well as the incompressibility equation (4.2). Denoting the weighting functions by $W_i(x, y)$ and $W_l^p(x, y)$ respectively, we obtain

$$
\left.
\begin{aligned}
\int_\Omega W_i\left[\mu\left(\frac{\partial^2 u}{\partial x^2} + \frac{\partial^2 u}{\partial y^2}\right) - \frac{\partial p}{\partial x}\right] dx\,dy = 0 \\
\int_\Omega W_i\left[\mu\left(\frac{\partial^2 v}{\partial x^2} + \frac{\partial^2 v}{\partial y^2}\right) - \frac{\partial p}{\partial y}\right] dx\,dy = 0
\end{aligned}
\right\}
\tag{4.4}
$$

and

$$
\int_\Omega W_l^p\left(\frac{\partial u}{\partial x} + \frac{\partial v}{\partial y}\right) dx\,dy = 0
\tag{4.5}
$$

Integration by parts of equation 4.4 yields for a domain Ω and boundary Γ

$$
\left.
\begin{aligned}
\int_\Omega\left[\mu\left(\frac{\partial W_i}{\partial x}\frac{\partial u}{\partial x} + \frac{\partial W_i}{\partial y}\frac{\partial u}{\partial y}\right) - \frac{\partial W_i}{\partial x}p\right] dx\,dy - \int_\Gamma W_i\bar{X}_1\,d\Gamma = 0 \\
\int_\Omega\left[\mu\left(\frac{\partial W_i}{\partial x}\frac{\partial v}{\partial x} + \frac{\partial W_i}{\partial y}\frac{\partial v}{\partial y}\right) - \frac{\partial W_i}{\partial y}p\right] dx\,dy - \int_\Gamma W_i\bar{Y}_1\,d\Gamma = 0
\end{aligned}
\right\}
\tag{4.6}
$$

with

$$
\bar{X}_1 = \mu\left(l_x\frac{\partial u}{\partial x} + l_y\frac{\partial u}{\partial y}\right) - l_x p = \mu\frac{\partial u}{\partial n} - l_x p
$$
$$
\bar{Y}_1 = \mu\left(l_x\frac{\partial v}{\partial x} + l_y\frac{\partial v}{\partial y}\right) - l_y p = \mu\frac{\partial v}{\partial n} - l_y p
\tag{4.7}
$$

where l_x and l_y represent the direction cosines of the normal n to surface element $d\Gamma$.

In order to obtain the approximate solution, we express the velocity $u(x, y)$, $v(x, y)$ and pressure $p(x, y)$ as

$$
u(x, y) = \sum N_i(x, y)u_i = \mathbf{Nu}, \qquad v(x, y) = \sum N_i(x, y)v_i = \mathbf{Nv}
\tag{4.8}
$$

and

$$
p(x, y) = \sum H_l(x, y)p_l = \mathbf{Hp}
\tag{4.9}
$$

where **u**, **v** and **p** represent velocity and pressure vectors whose components are nodal values u_i, v_i and p_l respectively. **N** and **H** are interpolation or shape functions. We adopt the Galerkin criterion in the selection of the weighting functions, i.e. we use the interpolation functions as the weighting functions

$$W_i(x, y) = N_i(x, y) \tag{4.10}$$

Equations 4.6 can now be written as

$$\left.\begin{aligned}
\int_\Omega \left[\mu\left(\frac{\partial N_i}{\partial x}\frac{\partial \mathbf{N}}{\partial x} + \frac{\partial N_i}{\partial y}\frac{\partial \mathbf{N}}{\partial y}\right)\mathbf{u} - \frac{\partial N_i}{\partial x}\mathbf{Hp} \right] dx\, dy - \int_\Gamma N_i \overline{X}_1\, d\Gamma = 0 \\
\int_\Omega \left[\mu\left(\frac{\partial N_i}{\partial x}\frac{\partial \mathbf{N}}{\partial x} + \frac{\partial N_i}{\partial y}\frac{\partial \mathbf{N}}{\partial y}\right)\mathbf{v} - \frac{\partial N_i}{\partial y}\mathbf{Hp} \right] dx\, dy - \int_\Gamma N_i \overline{Y}_1\, d\Gamma = 0
\end{aligned}\right\} \tag{4.11}$$

with

$$\overline{X}_1 = \mu\left(l_x\frac{\partial \mathbf{N}}{\partial x} + l_y\frac{\partial \mathbf{N}}{\partial y}\right)\mathbf{u} - l_x\mathbf{Hp}$$

$$\overline{Y}_1 = \mu\left(l_x\frac{\partial \mathbf{N}}{\partial x} + l_y\frac{\partial \mathbf{N}}{\partial y}\right)\mathbf{v} - l_y\mathbf{Hp} \tag{4.12}$$

A similar relation to Equation 4.11 can be written for Equation 4.5. We note that all above relations are obtainable from the stationary condition of the functional Π_1, defined below when Equations 4.8 and 4.9 are substituted

$$\Pi_1 = \int_\Omega \left\{ \tfrac{1}{2}\mu\left[\left(\frac{\partial u}{\partial x}\right)^2 + \left(\frac{\partial u}{\partial y}\right)^2 + \left(\frac{\partial v}{\partial x}\right)^2 + \left(\frac{\partial v}{\partial y}\right)^2\right] - p\left(\frac{\partial u}{\partial x} + \frac{\partial v}{\partial y}\right) \right\} dx\, dy$$

$$- \int_\Gamma (\overline{X}_1 u + \overline{Y}_1 v)\, d\Gamma \tag{4.13}$$

The 'stiffness' matrix which is derivable from Π_1 is obviously symmetric. Zienkiewicz[4] and Taylor and Hood [5] only used the integration by parts on the velocity terms in Equation 4.4 and consequently obtained an unsymmetrical form.*

Although we exclusively base our solution procedures on Equation 4.11 or the functional Π_1 of Equation 4.13, some remarks on alternative formulations would be in order. In view of the incompressibility relation (4.2), the Stokes equation (4.3) can be written in alternate forms as follows:

$$2\mu\left[\frac{\partial^2 u}{\partial x^2} + \frac{1}{2}\frac{\partial}{\partial y}\left(\frac{\partial u}{\partial y} + \frac{\partial v}{\partial x}\right)\right] - \frac{\partial p}{\partial x} = 0$$

$$2\mu\left[\frac{\partial^2 v}{\partial y^2} + \frac{1}{2}\frac{\partial}{\partial x}\left(\frac{\partial u}{\partial y} + \frac{\partial v}{\partial x}\right)\right] - \frac{\partial p}{\partial y} = 0 \tag{4.14}$$

---

* In the first printing of Reference 4, Zienkiewicz states that a functional can not exist. This error is corrected in subsequent printings.

or

$$2\mu\left[\frac{1}{3}\frac{\partial}{\partial x}\left(2\frac{\partial u}{\partial x} + \frac{\partial v}{\partial y}\right) + \frac{1}{2}\frac{\partial}{\partial y}\left(\frac{\partial u}{\partial y} + \frac{\partial v}{\partial x}\right)\right] - \frac{\partial p}{\partial x} = 0$$

$$2\mu\left[\frac{1}{2}\frac{\partial}{\partial x}\left(\frac{\partial u}{\partial y} + \frac{\partial v}{\partial x}\right) + \frac{1}{3}\left(-\frac{\partial u}{\partial x} + \frac{\partial v}{\partial y}\right)\right] - \frac{\partial p}{\partial y} = 0$$

(4.15)

Equation 4.14 can be obtained from the stationary condition of the following functional $\Pi_2{}^*$:

$$\Pi_2 = \int_\Omega \left\{ \mu\left[\left(\frac{\partial u}{\partial x}\right)^2 + \left(\frac{\partial v}{\partial y}\right)^2 + \frac{1}{2}\left(\frac{\partial u}{\partial y} + \frac{\partial v}{\partial x}\right)^2\right] - p\left(\frac{\partial u}{\partial x} + \frac{\partial v}{\partial y}\right) \right\} dx\,dy$$

$$- \int_\Gamma (\bar{X}_2 u + \bar{Y}_2 v)\,d\Gamma$$

(4.16)

where

$$\bar{X}_2 = 2\mu\left[l_x\frac{\partial u}{\partial x} + \tfrac{1}{2}l_y\left(\frac{\partial u}{\partial y} + \frac{\partial v}{\partial x}\right)\right] - l_x p$$

$$\bar{Y}_2 = 2\mu\left[\tfrac{1}{2}l_x\left(\frac{\partial u}{\partial y} + \frac{\partial v}{\partial x}\right) + l_y\frac{\partial v}{\partial y}\right] - l_y p$$

(4.17)

The functional Π_2 is described by Tong[12] and is analogous to the one that Herrmann[13] proposed as a means for treating the plane strain problems of incompressible elastic solids. $\bar{X}_2$ and $\bar{Y}_2$ of Equation 4.17 are the surface tractions.[5,6,14] It is, however, difficult to envisage the natural boundary conditions which can be specified by such tractions in fluid flow problem.

Finally, Equation 4.17 can be derived from the functional

$$\Pi_3 = \int_\Omega \left\{ \tfrac{2}{3}\mu\left[\left(\frac{\partial u}{\partial x}\right)^2 + \left(\frac{\partial v}{\partial y}\right)^2 - \frac{\partial u}{\partial x}\frac{\partial v}{\partial y}\right] + \tfrac{1}{2}\mu\left(\frac{\partial u}{\partial y} + \frac{\partial v}{\partial x}\right)^2 \right.$$

$$\left. - p\left(\frac{\partial u}{\partial x} + \frac{\partial v}{\partial y}\right) \right\} dx\,dy - \int_\Gamma (\bar{X}_3 u + \bar{Y}_3 v)\,d\Gamma$$

(4.18)

with

$$\bar{X}_3 = 2\mu\left[\tfrac{1}{3}l_x\left(2\frac{\partial u}{\partial x} - \frac{\partial v}{\partial y}\right) + \tfrac{1}{2}l_y\left(\frac{\partial u}{\partial y} + \frac{\partial v}{\partial x}\right)\right] - l_x p$$

$$\bar{Y}_3 = 2\mu\left[\tfrac{1}{2}l_x\left(\frac{\partial u}{\partial y} + \frac{\partial v}{\partial x}\right) + \tfrac{1}{3}l_y\left(2\frac{\partial v}{\partial y} - \frac{\partial u}{\partial x}\right)\right] - l_y p$$

(4.19)

---

* This functional is identical to the one derived in Chapter 11 (Equation 2.16 with constraint of Equation 2.16a).

4.2.2   Solution procedure

We use the triangular element of Figure 4.1 and adopt a quadratic interpolation for velocity components u, v and a linear one for fluid pressure p. The functions N_i and H_i in Equations 4.8 and 4.9 can be expressed, in terms

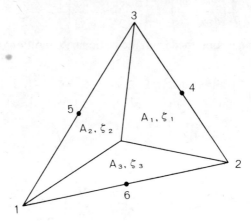

Figure 4.1   Triangular element with corner nodes 1, 2, 3 and mid-edge nodes 4, 5, 6, ζ_1, ζ_2 and ζ_3 are area coordinates

of the area coordinates L_1, L_2 and L_3, as :

$$\left.\begin{aligned}
N_1 &= L_1^2 - L_1(L_2 + L_3), & N_2 &= L_2^2 - L_2(L_3 + L_1), \\
N_3 &= L_3^2 - L_3(L_1 + L_2), & N_4 &= 4L_2L_3, \\
N_5 &= 4L_3L_1, & N_6 &= 4L_1L_2
\end{aligned}\right\} \qquad (4.20)$$

and

$$H_1 = L_1, \qquad H_2 = L_2, \qquad H_3 = L_3 \qquad (4.21)$$

Equation 4.9 is written as

$$p(x, y) = [L_1 \ L_2 \ L_3] \{\mathbf{p}\} \qquad (4.22)$$

From Equation 4.5 we have

$$\int_\Omega W_i \left(\frac{\partial \lfloor N \rfloor}{\partial x} \{u\} + \frac{\partial \lfloor N \rfloor}{\partial y} \{v\} \right) dx \, dy = 0 \qquad (4.23)$$

In our formulation, the integration by parts is not applied to the incompressibility relation given by Equation 4.23.

One reason for the choice of interpolation functions for the velocity components which are one order higher than those of the fluid pressure is that the associated integral of Equation 4.11 contains derivatives of the velocity field (first derivatives) that are one order higher than those of the pressure field (zeroth order derivatives). Furthermore, as will be demonstrated below, the algebraic equations resulting from a linear velocity field may be over-constrained by the corresponding linear pressure field.

The stiffness equation derived from Equation 4.11, together with Equation 4.23, assumes the following form at the element level

$$\mathbf{K}\boldsymbol{\psi} = \mathbf{f} \qquad (4.24)$$

or

$$
\begin{bmatrix}
\mathbf{k}^u & \mathbf{0} & \mathbf{k}^{up} \\
\mathbf{0} & \mathbf{k}^v & \mathbf{k}^{vp} \\
\mathbf{k}^{pu} & \mathbf{k}^{pv} & \mathbf{0}
\end{bmatrix}
\begin{Bmatrix}
u_1 \\ \vdots \\ u_6 \\ \hline v_1 \\ \vdots \\ v_6 \\ \hline p_1 \\ \vdots \\ p_3
\end{Bmatrix}
=
\begin{Bmatrix}
f_1 \\ \vdots \\ f_6 \\ \hline f_7 \\ \vdots \\ f_{12} \\ \hline 0
\end{Bmatrix}
$$

The 6×6 matrix $[\mathbf{k}^u]$ is symmetric. When we base our computation on Equation 4.11, the matrices $\mathbf{k}^{up}$ and $\mathbf{k}^{vp}$ are related to $\mathbf{k}^{pu}$ and $\mathbf{k}^{pv}$ as

$$\mathbf{k}^{pu} = -\mathbf{k}^{up\,\mathrm{T}}, \qquad \mathbf{k}^{vp} = -\mathbf{k}^{pv\,\mathrm{T}} \qquad (4.25)$$

We can thus make the matrix $\mathbf{k}$ symmetric by replacing the force vector $\mathbf{p}$ by $-\mathbf{p}$.

The assembled overall stiffness equation can be written as

$$
\begin{bmatrix}
\mathbf{K}_{aa} & \mathbf{K}_{ar} \\
\mathbf{K}_{ra} & \mathbf{K}_{rr}
\end{bmatrix}
\begin{Bmatrix}
\boldsymbol{\Psi}_a \\ \boldsymbol{\Psi}_r
\end{Bmatrix}
=
\begin{Bmatrix}
\mathbf{P}_a \\ \mathbf{P}_r
\end{Bmatrix}
\qquad (4.26)
$$

and is solved for the unknown vector $\boldsymbol{\Psi}_a$, writing

$$\mathbf{K}_{aa}\boldsymbol{\Psi}_a = \mathbf{P}_a - \mathbf{K}_{ar}\boldsymbol{\Psi}_r \qquad (4.27)$$

where $\boldsymbol{\Psi}_r$ represents the nodal values of prescribed velocity and/or fluid pressure along the boundary. It should be noticed that $\mathbf{K}_{aa}\boldsymbol{\Psi}_a$ has the form

$$
\begin{array}{c}
 \quad m \qquad n \\
\begin{array}{c} m \\ n \end{array}
\begin{bmatrix}
\mathbf{K}_{aa} & \mathbf{K}_{ab} \\
\mathbf{K}_{ba} & \mathbf{0}
\end{bmatrix}
\begin{Bmatrix}
\mathbf{u} \\ \mathbf{p}
\end{Bmatrix}
\end{array}
\qquad (4.28)
$$

where **u** contains all node point values of u_i and v_i and **p** lists the node point pressures. It can be deduced from Equation 4.28 that the number of components m of unknown velocity vector **u** should not be less than the number of components n of unknown pressure vector **p**, i.e.

$$m \geqslant n \qquad (4.29)$$

In usual fluid flow problem, most of the boundary conditions are prescribed in terms of velocity. Accordingly, the fulfilment of Equation 4.29 is a hard task in some cases, particularly when we adopt an interpolation function of the same order for both velocity components and fluid pressure. This also justifies our choice of different types of interpolation functions given by Equations 4.20 and 4.21.

As for boundary conditions, we can classify the boundary into two categories Γ_1 and Γ_2. On Γ_1, both components of velocity are prescribed. On the remaining boundary Γ_2, we specify X_1 and X_2 of Equation 4.7. As is done in the solid continuum, the sum of components of $\mathbf{P}_a$ of Equation 4.26 at an internal node is equated to zero.

4.2.3 Numerical results of Stokes flow

Using the finite element mesh of Figure 4.2, the Couette flow between parallel plates has been studied. The boundary conditions are given as follows:

constant values of p, $\partial u/\partial n$ ($= 0$) and $\partial v/\partial n$ ($= 0$) on AB and DC

constant values of u and v ($= 0$)   on BC and AD

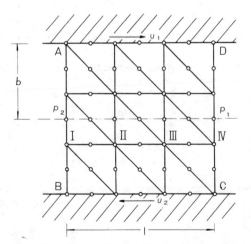

Figure 4.2 Finite element mesh used for analysis of Couette flow;
$b = 1$, $l = 2\,\text{cm}$, $u_1 = 1\cdot 0$, $u_2 = -0\cdot 5\,\text{cm/s}$
$p_1 = 0\cdot 0$, $p_2 = 54\cdot 0\,\text{g/s}^2\,.\,\text{cm}$, $\mu = 13\cdot 5\,\text{gr/s}\,.\,\text{cm}$

It can be seen from Figure 4.3 that the present numerical results agree completely with the theoretical solutions. The horizontal velocity distribution u at every section x = constant is identical and the vertical velocity v was found to be practically zero in the numerical solution. Further, a linear drop of fluid pressure p in the direction of x is observed.

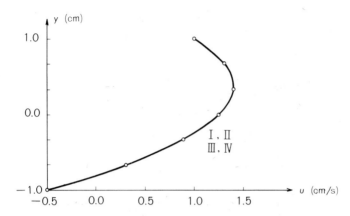

Figure 4.3   Distribution of horizontal velocity (Couette flow)

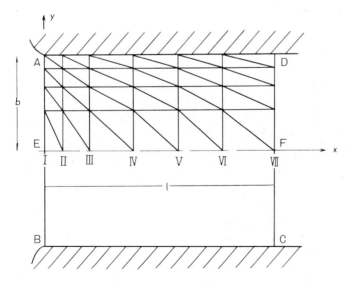

Figure 4.4   Finite element mesh used for analysis of entry flow;
$b = 1$, $l = 2.6$ cm

The second example is concerned with an entrance flow. The finite
element subdivision is shown in Figure 4.4 and the boundary conditions are

$$u = 2/3 \quad \text{and} \quad v = 0 \qquad \qquad \text{on AE}$$

$$u = 0 \quad \text{and} \quad v = 0. \qquad \qquad \text{on AD}$$

$$\partial u/\partial n = 0, \quad \partial v/\partial n = 0 \quad \text{and} \quad p = 0 \quad \text{on DF}$$

$$v = 0, \quad \partial u/\partial n = 0 \quad \text{and} \quad pl_x = 0 \quad \text{on EF}$$

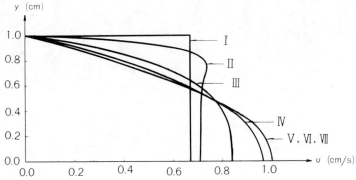

Figure 4.5   Transition of uniform velocity at the entry to Poiseuille parabolic flow

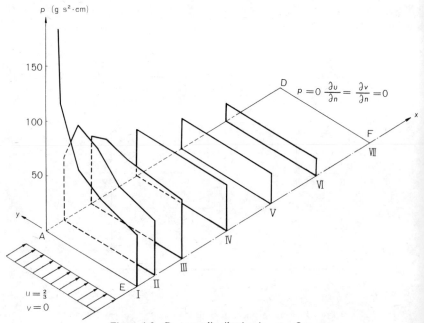

Figure 4.6   Pressure distribution in entry flow

Figure 4.5 shows the transition of uniform horizontal entry velocity to two-dimensional Poiseuille flow along the x axis. The pressure distribution is illustrated in Figures 4.7 and 4.8. The solutions are in reasonably close qualitative and quantitative agreement with those obtained by Atkinson *et al.*[3]

4.2.4 Iteration for Navier–Stokes equation

To extend the solution of the Stokes equation (4.3) in the preceding section to the full Navier–Stokes equation (4.1) with convective terms, we employ the following iterative procedure.

Denoting an approximate solution by u^*, v^* and p^*, we introduce this into the coefficient of convective terms of Equation 4.1. Thus, we have

$$\left. \begin{aligned} \mu\left(\frac{\partial^2 u}{\partial x^2} + \frac{\partial^2 u}{\partial y^2}\right) - \rho\left(u^*\frac{\partial u}{\partial x} - v^*\frac{\partial u}{\partial y}\right) - \frac{\partial p}{\partial x} = 0 \\ \mu\left(\frac{\partial^2 v}{\partial x^2} + \frac{\partial^2 v}{\partial y^2}\right) - \rho\left(u^*\frac{\partial v}{\partial x} - v^*\frac{\partial v}{\partial y}\right) - \frac{\partial p}{\partial y} = 0 \end{aligned} \right\}$$

(4.30)

Generally, u^*, v^* and p^* represent the nth iteration values of the Navier–Stokes equation. At start, these may be the solution of the Stokes equation or the one for a lower value of Reynolds number. A system of algebraic equations can be obtained by employing the weighted residual method as in Equation 4.4. To keep the natural boundary conditions identical with those for the Stokes equation, the integration by parts is not applied to the convective terms. The resulting stiffness equation is expressed as

$$\mathbf{K}^*\mathbf{\Psi} = \mathbf{f} \tag{4.31}$$

In contrast to Equation 4.24 for the Stokes equation, the matrix $\mathbf{K}^*$ in Equation 4.31 is no longer symmetric.

We solve Equation 4.31 by modifying successively the values of u^* and v^* in $\mathbf{K}^*$ until the solution for u, v and p converges to a limiting one with tolerable error. We have set the error criterion as

$$\max\left(\left|\frac{u_i^{n+1} - u_i^n}{u_i^{n+1}}\right|, \left|\frac{v_i^{n+1} - v_i^n}{v_i^{n+1}}\right|, \left|\frac{p_l^{n+1} - p_l^n}{p_l^{n+1}}\right|\right) < 10^{-6} \tag{4.32}$$

where superscripts n and $n + 1$ indicate the successive stages of iteration. Incorporation of the over-relaxation technique is advisable to accelerate the speed of convergence. It is worth emphasizing that we do not treat the convective terms as the 'apparent' forces, but incorporate their contributions into the 'stiffness' matrix $\mathbf{K}^*$ of Equation 4.31. A similar method has been adopted by Taylor and Hood.[5]

As a numerical example, the flow past a circular cylinder is analysed. As depicted in Figure 4.7, the infinite field of flow is confined by a circle with radius $r/a = 8.0$ or 20.0 and divided into finite element regions. The external

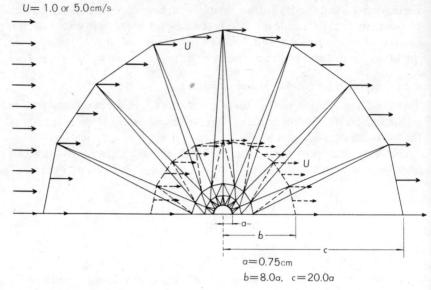

Figure 4.7 Boundary conditions and finite element mesh for the analysis of the flow past a
circular cylinder

boundary conditions are specified at these radii as

$$U = 5.0, \qquad V = 0 \qquad \text{and} \qquad p = 0 \quad \text{for Reynolds number } Re = 30.0$$

Boundary conditions on the surface $r = a$ of the cylinder are simply

$$U = 0 \qquad \text{and} \qquad V = 0$$

Figure 4.8 shows the velocity distribution obtained for the Stokes equation.

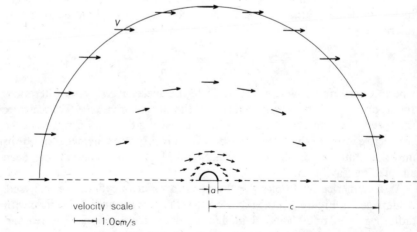

Figure 4.8 Velocity solution of the Stokes equation

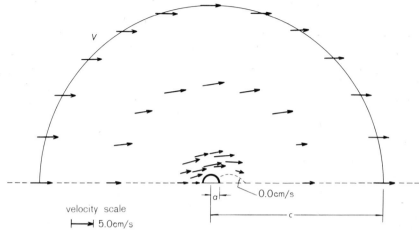

Figure 4.9   Velocity distribution for Reynolds number $= 30 \cdot 0$

The velocity distribution becomes unsymmetric for the Navier–Stokes flow, due to the inclusion of convected terms. Figure 4.9 gives the velocity distribution for $Re = 30$. This is reached by eight iterations from the Stokes solution.

Figure 4.10 shows the pressure distribution around the cylinder. The point to note is that the solution of the Stokes flow is very sensitive to the location of the boundary laid down at the beginning. This corresponds to

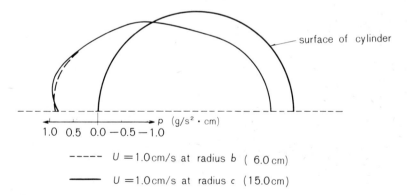

Figure 4.10   Pressure distribution; solution of the Navier–Stokes equation for flow of Reynolds number $= 30 \cdot 0$

the theoretical prediction that the Stokes equation or its expansion solution cannot satisfy the full boundary condition at infinity.[15,16] In view of this, the direct numerical solution of the Navier–Stokes equation as described in the present paper is considered to be satisfactory.

4.3 Metal steady flow

Before the introduction of elastic-plastic analysis procedures by the finite element method, most plasticity problems were approached by the plastic-rigid slip-line field theory. Steady flow of metals is not an exception and it is rather surprising that the slip-line theory was successful in providing many useful solutions for continuous shaping processes such as sheet-drawing and extrusion.[17] The uniqueness of stress solution in the plastically deforming region has been well established[18] for perfectly plastic solids and there have been a number of examples that suggest the plastic-rigid solutions are physically feasible ones.

Although the 'full collapse load' state of perfectly plastic solids can be reached by the elastic-plastic finite element solution procedure, it is desirable to establish a method that can analyse the steady mode of deformation separately. The existing methods are not able to incorporate the history of deformation, e.g. the strain-hardening of the material element along the path followed. Moreover, there have been a variety of problems that have not been given the 'complete' solution in the sense of the theory of perfectly plastic-rigid solids. The metal strip rolling is such an example. A new method of solution is developed in the present study. It is summarized as follows.

(1) The elastic and unsteady elastic-plastic problem is solved by the finite element method.[10,11] For the solution, the initial rigid geometric configuration of the metal being worked may be used. It is desirable, however, to start the calculation by assuming stream lines and consequently the deformed shape of the metal. The mesh subdivision in this case is based on the assumed stream lines. An example of assumed stream line and of the finite element mesh is shown in Figure 4.11.

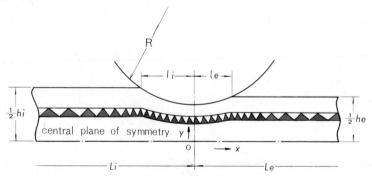

Figure 4.11 Assumed streamline and finite element mesh. Rolling problem

The solution at this step provides the first approximation for the nodal displacement, strain- and stress-increments which the material element experiences in the path of the steady mode of deformation.

(2) Using the increments obtained in step 1 and integrating along the path (see Figure 4.12), determine the mechanical states of material element, i.e. stress, strain and equivalent stress. The equivalent stress is the measure for

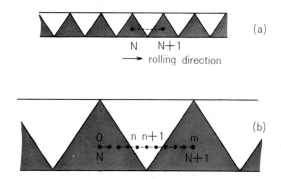

Figure 4.12   Integration path along a line (approximate streamline) passing through centroid of triangular element

distinguishing whether the material is in the elastic, plastic or unloading state. In the example described below, the following integration scheme is used

$$df_{N\sim(N+1)} = \frac{A_N\,df_N + A_{N+1}\,df_{N+1}}{A_N\,d\bar{x}_N + A_{N+1}\,d\bar{x}_{N+1}}(\bar{x}_{N+1} - \bar{x}_N) \qquad (4.33)$$

where df represents the increment of the relevant quantity. Subscripts N and $N + 1$ indicate the successive adjacent triangular elements. A, $\bar{x}$ and $d\bar{x}$ with subscripts N and $N + 1$ are the area, the x coordinate and the incremental displacement of the centroid of triangles N and $N + 1$. Note that the integration is carried out among the same class of elements, e.g. shaded ones $1, \ldots N - 1$, N, $N + 1, \ldots$ in the example under study. To improve the solution, it is sometimes advantageous to subdivide the integration path into m segments as shown in Figure 4.12(b).

Nodal forces are determined from the calculated stress in the element. The formula for computing nodal forces $\mathbf{F}$, written for the generic Nth element, is

$$\mathbf{F}_N = A_N \mathbf{B}_N^T \boldsymbol{\sigma}_N \qquad (4.34)$$

where A_N and $\boldsymbol{\sigma}_N$ represent the area of the triangle and the calculated stress at the centroid. $\mathbf{B}$ is the usual strain-displacement matrix and it will be noticed here that the simplex triangular element is adopted in our calculation. The $\mathbf{F}$ of Equation 4.34 is used for obtaining the contact stress at nodes along the tool-work interface as well as for checking the force equilibrium at internal nodes.

(3) Modify the stress–strain matrix **D** pertaining to each finite element depending on the mechanical states defined in step (2). Modification of stream lines as well as the adjustment of the extent of contact area between tool and work are advisable at this stage, if necessary. When the calculated normal force is tensile at nodal points located along the tool–work interface, we regard the points as not being in contact. Conversely, when the relative displacement between two opposite nodal points, one on tool face and the other on material surface, exceeds the gap between them, we assume they are in a contact state for the subsequent computation.

(4) By carrying out the finite element analysis as in step (1), we can obtain an improved solution for the nodal displacement and strain- and stress-increments.

(5) Repeat the steps (2) to (4), until the difference between successive approximations becomes less than the nominal accuracy of calculation.

Our first successful example was the skin-pass rolling of sheet strip. The relevant geometric data are summarized in Table 4.1. The thickness

Table 4.1   Geometric data in example problem and calculated roll forces (material of working roll: rigid)

		Elastic case	Elastic-plastic
Radius of roll	R (mm)	1000·0	250·0
Thickness before rolling	h_i (mm)	2·0000	2·0000
Thickness after rolling	h_e (mm)	2·0000	1·9946
Thickness reduction	r (%)	0·00	0·27
Contact length in entry side	l_i (mm)	2·0000	1·6000
Contact length in exit side	l_e (mm)	2·0000	1·1000
Sheet length in entry side	L_i (mm)	5·785	5·616
Sheet length in exit side	L_e (mm)	5·785	4·552
Number of nodal points		11 × 76	11 × 76
Number of elements		20 × 75	20 × 75
Calculated roll force	P (kg/mm)	168·3	175·5

reduction or the overall deformation of the material in this process is rather small and it was found that the incorporation of a correction for geometric non-linearities is not essential. Figure 4.13 shows the finite element mesh division used and the deformed metal configuration together with the contact pressure distribution. Figure 4.14 depicts the calculated stress distribution in the surface and central regions of the strip for elastic solution. The elastic material constants are: Young's modulus $E = 2 \cdot 1 \times 10^4$ kg/mm^2 and Poisson ratio $\nu = 0 \cdot 30$. It can be seen that the strip deformation, the normal stress σ_x, σ_y, as well as the roll pressure distribution, are symmetric in this elastic case where the boundary condition is specified so that the resultant horizontal forces are zero in both exit and entry planes at a

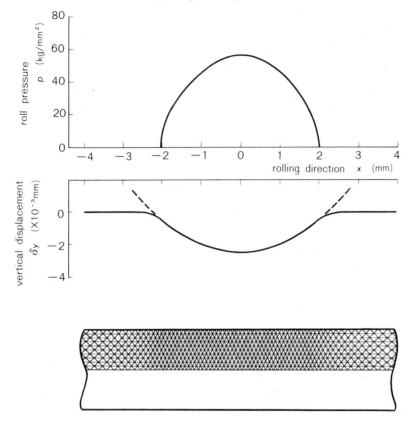

Figure 4.13   Contact pressure distribution in elastic case and finite element mesh

sufficiently great distance from the roll. Consistent with above, the shear stress τ_{xy} is antisymmetric with zero value at the roll centre. Sticking frictional condition is assumed in the example of the present paper so that the metal has the same velocity as the roll periphery along the arc of contact.

Figure 4.15 illustrates the deformed configuration and contact pressure distribution when the plasticity of the relevant material is taken into consideration. In the computation, the following strain-hardening law is assumed:

$$\bar{\sigma} = 100 \cdot 0 (0 \cdot 0625 + \bar{\varepsilon}^p)^{0 \cdot 25} \tag{4.35}$$

where $\bar{\sigma}$ and $\bar{\varepsilon}^p$ are the equivalent stress and the equivalent plastic strain respectively. The yield stress is given by $Y = 100 \cdot 0 \times (0 \cdot 0625)^{0 \cdot 25} = 50$ kg/mm². Figure 4.16 shows the stress distribution along the strip surface as well as the central region. It is rather striking that there exist intense

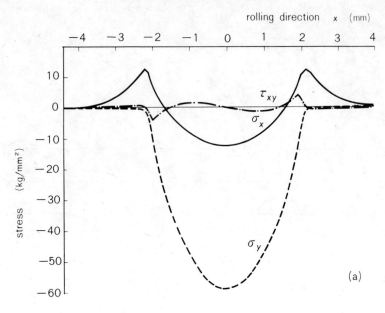

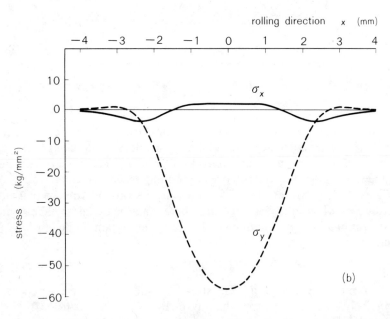

Figure 4.14 Stress distribution in (a) surface layer and (b) central plane $y = 0$ of symmetry; elastic case

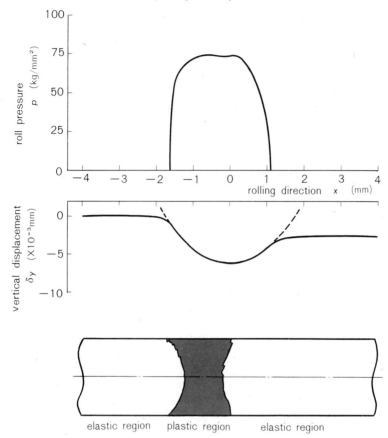

Figure 4.15 Contact pressure in actual skin-pass process giving plastic deformation to metal strip

shear zones around entry and exit regions, even in the case of skin-pass rolling with rather small thickness reduction. Although the present solution discloses this kind of shear zone, the numerical results are prone to be affected considerably by the presence of the intense shear, when we carry out integration along the material path in order to assess the change of mechanical state. The deterioration of numerical accuracy is vital when we extend our computation to problems with large reduction of thickness or area. Therefore, the use of quadratic element and/or higher order polynomial interpolation is considered to be mandatory in the analysis of high reduction continuous processes such as sheet drawing and extrusion.

In concluding, we observe that the application of the fluid flow theory of the first part of the present paper to material processing analysis might merit

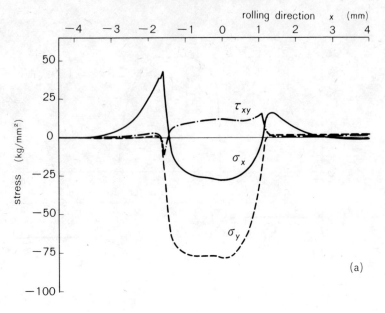

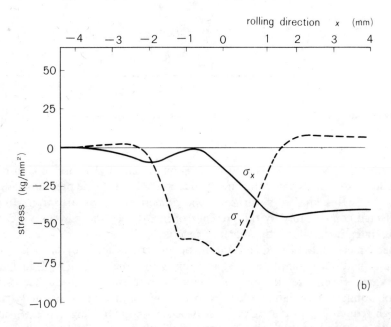

Figure 4.16 Stress distribution in (a) surface layer and (b) central axis $y = 0$ of symmetry; plastic case

investigation. Such an attempt has been made in the past in a comparison of the conventional rolling theories with the 'shear-plane' theory.[19] It must be emphasized, however, that the finite element method facilitates the comparison by providing detailed and accurate numerical data and thus greatly helps our physical reasoning of the problems under study.

Acknowledgements

The authors wish to acknowledge Professor F. Naruse, Institute of Industrial Science, for his kind advice and discussion which were very helpful in deepening their understanding of the theory of fluid mechanics. Special thanks also go to Professor R. H. Gallagher, who spent the autumn of 1973 in Japan, for reviewing the manuscript of the present paper and making many useful suggestions.

References

1. M. D. Olson, 'A variational finite element method for two-dimensional steady viscous flows', *Proceedings of the Speciality Conference on Finite Element Method in Civil Engineering*, McGill University, 585–616 (1972).
2. O. C. Zienkiewicz and C. Taylor, 'Weighted residual processes in finite element with particular reference to some transient and coupled problems', *Lectures on Finite Element Methods in Continuum Mechanics* (eds. J. T. Oden and E. R. A. Oliveira), University of Alabama Press, 415–458 (1973).
3. B. Atkinson, M. P. Brocklebank, C. C. H. Card and J. M. Smith, 'Low Reynolds number developing flows', *AIChE Journal*, **15**, 4, 548–553 (1969).
4. O. C. Zienkiewicz, *The Finite Element Method in Engineering Science*, Chapters 3 and 15, McGraw-Hill, London, 1971.
5. C. Taylor and P. Hood, 'A numerical solution of the Navier–Stokes equations using the finite element technique', *Computers and Fluids*, **1**, 73–100 (1973).
6. M. Kawahara, N. Yoshimura, K. Nakagawa *et al.*, 'Steady flow analysis of incompressible viscous fluid by the finite element method', *Theory and Practice in Finite Element Structural Analysis* (eds. Y. Yamada and R. H. Gallagher), University of Tokyo Press, 557–572 (1973).
7. I. F. Collins, 'The algebraic-geometry of slip line fields with applications to boundary value problems', *Proc. Roy. Soc.*, **A303**, 317–338 (1968).
8. I. F. Collins, 'Compression of a rigid-perfectly plastic strip between parallel rotating smooth dies', *Quart. Journ. Mech. and Applied Math.*, **23**, 329–348 (1970).
9. P. Dewhurt and I. F. Collins, 'A matrix technique for constructing slip-line field solutions to a class of plane strain plasticity problems', *Int. J. for Numerical Methods in Engineering*, **7**, 357–378 (1973).
10. Y. Yamada, N. Yoshimura and T. Sakurai, 'Plastic stress–strain matrix and its application for the solution of elastic-plastic problems by the finite element method', *Int. J. Mechanical Science*, **10**, 343–354 (1968).
11. P. V. Marcal, 'Finite element analysis with material nonlinearities—theory and practice', *Recent Advances in Matrix Methods of Structural Analysis and Design* (eds. R. H. Gallagher *et al.*), University of Alabama Press, 257–282 (1971).

12. P. Tong, 'The finite element method for fluid flow', *Recent Advances in Matrix Methods of Structural Analysis and Design* (eds. R. H. Gallagher *et al.*), University of Alabama Press, 787–808 (1971).
13. L. R. Herrmann, 'Elasticity equations for incompressible or nearly incompressible materials by a variational theorem', *AIAA Journal*, **3**, 1896–1900 (1965).
14. J. T. Oden, 'The finite element method in fluid mechanics', *Lectures on Finite Element Methods in Continuum Mechanics* (eds. J. T. Oden and E. R. A. Oliveira), University of Alabama Press, 151–186 (1973).
15. I. Proudman and J. R. A. Pearson, 'Expansion at small Reynolds numbers for the flow past a sphere and a circular cylinder', *Journal of Fluid Mechanics*, **2**, 237–262 (1957).
16. J. D. Cole, *Perturbation Methods in Applied Mechanics*, Blaisdell, 1968.
17. R. Hill, *The Mathematical Theory of Plasticity*, Oxford University Press, 1950.
18. R. Hill, 'On the state of stress in a plastic-rigid body at the yield point', *Phil. Mag.*, **42**, 868–875 (1951).
19. I. L. May and K. D. Nair, 'On the relative validity of two current cold-rolling theories', *Journal of Basic Engineering, Transactions ASME*, 69–75 (March, 1967).

Chapter 5

Tidal Propagation and Dispersion in Estuaries

C. Taylor and J. M. Davis

5.1 Introduction

Numerous theories have been advocated in an attempt to provide tractable solutions to both the wave theory governing tidal propagation and dispersion in estuaries. With the advent of digital computers, rational attempts have been made to include prototype geometric complexity, equation non-linearity and the dispersive mechanism in a piecewise manner. These are, of necessity, numerical and use the well known finite difference[1,2,3,4,5,6] or characteristic techniques.[7,8] These techniques are well documented in the indicated texts and will not be discussed in detail. In each, the solution progresses, in a stepwise manner, from some prescribed initial conditions. However, both are subject to the usual difficulties encountered when complicated geometric configurations and boundary conditions exist.

Although the simulation of wave motion within a tidal domain is in itself difficult the prediction of effluent dispersion is a formidable task. This difficulty is not associated with the mathematical or numerical description of the physical domain but in the definition of the mechanics of dispersion and diffusion. If such uncertainties are associated with the possible biological and chemical complexities of each effluent species, either singly or collectively, then the task of accurate assessment becomes formidable. The application of any numerical method to evolve a solution to the equations governing the above phenomena is, therefore, sometimes restricted by the absence of meaningful data relating to the necessary physical, biological and chemical processes within the ecosystem. The purpose of this chapter is to present a numerical method which can be used to formulate a model of both tidal propagation and dispersion in estuaries, rivers and seas. This model, like any other, is subject to limitations associated with the definition of the parameters necessary for solution as indicated above.

In order to achieve a meaningful analysis the numerical model should simulate the prototype current velocity, water-depth and effluent dispersion, both spatially and temporally, during a tidal cycle. In addition, to complete the model, any biological, chemical and decay processes should be coupled

95

to the physical dispersive mechanism. The relevant equations of continuity, momentum and dispersion can be combined to set up a numerical model of the physical dispersion for a particular domain of known geometry. The validity of the model should, however, be checked by comparison with prototype measurements before any extrapolation, or indeed interpolation, of known data can be made. A quantitative measure of the distribution of both water velocity and depth is, owing to the temporal nature of the required parameters, particularly difficult under prototype conditions. However, postulations relating to a 'typical' tide can be made with reasonable accuracy. Once the model has been checked, extreme values such as the coincidence of maximum effluent input and low tidal range, is then simply a particular example. The magnitude of dispersion coefficients, obtained by using suitable tracers in either the prototype or a physical model, are again both spatial and temporal, and typical values are calculated.

The most difficult to quantify, at the present time, are any biological or chemical processes associated with a particular effluent when either considered in isolation or in association with other effluents within the domain. For instance the decomposition of sewage, measurable by the biochemical oxygen demand, is a temporal function of effluent concentration, ambient conditions, bacterial population, the toxic level of other effluents to bacterial metabolism, generation and growth. These highly complex interrelationships are currently the subject of a considerable amount of fundamental research. However, the availability of quantitative information is noticeably lacking.

Obviously the introduction of such complexities into even a problem of simple geometry would lead to analytically intractable equations. Therefore, recourse has usually been made either to numerical techniques or to physical model studies in an attempt to obtain a rational basis for design. Owing to the advent of large capacity digital computers it has been possible to achieve both reasonable accuracy and economy by utilizing numerical models. A distinct advantage of such models is the speed with which an analysis can be conducted for any known change in input or sanitary conditions. This means that for a maintained effluent outfall distribution, into a tidal domain, the immediate and long-term effects of any undesirable overload can be quickly analysed. The model is, therefore, a particularly useful tool for system management and control.

The authors present one such numerical model, based on the finite element method, and test its validity against values obtained from known analytical and other numerical solutions.

5.2 The basic equations

The three fundamental sets of equations embodied in the present study are those associated with the tidal hydraulics, dispersion and effluent reaction description within the polluted environment. All have been referred to in previous texts, typically,[1,9] the essential stages in their transformation into a form suitable for discretization will be outlined in the present section.

5.2.1 Tidal motion

The set of partial differential equations that describe the conservation of momentum and continuity in an Eulerian system[1] are

$$\frac{\partial u}{\partial t} + u\frac{\partial u}{\partial x} + v\frac{\partial u}{\partial y} + w\frac{\partial u}{\partial z} + \frac{1}{\rho}\frac{\partial p}{\partial x} = X \qquad (5.1)$$

$$\frac{\partial v}{\partial t} + u\frac{\partial v}{\partial x} + v\frac{\partial v}{\partial y} + w\frac{\partial v}{\partial z} + \frac{1}{\rho}\frac{\partial p}{\partial y} = Y \qquad (5.2)$$

$$\frac{\partial w}{\partial t} + u\frac{\partial w}{\partial x} + v\frac{\partial w}{\partial y} + w\frac{\partial w}{\partial z} + \frac{1}{\rho}\frac{\partial p}{\partial z} = Z \qquad (5.3)$$

and the continuity condition for an incompressible flow,

$$\frac{\partial u}{\partial x} + \frac{\partial v}{\partial y} + \frac{\partial w}{\partial z} = 0 \qquad (5.4)$$

where x, y, z describe the cartesian coordinates, Figure 5.1, with the z axis vertical and the x, y plane is horizontal and coincident with the mean sea

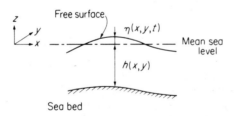

Figure 5.1   Vertical section

level, u, v, w are the components of velocity in the x, y and z directions respectively; p is the pressure; ρ is the density and X, Y, Z are components of extraneous force per unit mass. These include friction forces at the surface and bed, the Coriolis effect and the gravity force in the z direction.

Provided suitable expressions for the functions X, Y, Z are defined in this system then no difficulty arises in formulating a three-dimensional algorithm. However, the propagation of long period waves is essentially two-dimensional, which leads to simpler algebra and computational economy both in time and storage.

For the purposes of this chapter the general three-dimensional theory presented above will be reduced to a two-dimensional system via the long-period wave theory and definitions used by Dronkers[2] and Leendertse.[3]

Replacing atmospheric pressure term by an equivalent head h' and omitting tide generating forces (see Leendertse[3]), the dynamic equations can be written

$$\frac{\partial U}{\partial t} + U\frac{\partial U}{\partial x} + V\frac{\partial U}{\partial y} - \omega V = -g\frac{\partial}{\partial x}(h - h') - g\frac{\Upsilon . U}{C_C^2(H + h)} + \frac{\tau_x}{(H + h)}$$

$$(5.5)$$

$$\frac{\partial V}{\partial t} + U\frac{\partial V}{\partial x} + V\frac{\partial V}{\partial y} + \omega U = -g\frac{\partial}{\partial y}(h - h') - gC_C^2\frac{\Upsilon . V}{(H + h)} + \frac{\tau_y}{(H + h)} \qquad (5.6)$$

for the section shown on Figure 5.1 and typically

$$U = \frac{1}{(H + h)}\int_{-H}^{h} u\,dz \qquad (5.7)$$

$$V = \frac{1}{(H + h)}\int_{-H}^{h} v\,dz \qquad (5.8)$$

the vertically integrated velocity and

$$\Upsilon = (U^2 + V^2)^{\frac{1}{2}}$$

C_C = the Chézy friction coefficient

H = depth measured from the mean sea level

τ_x, τ_y are the components of the wind force on the water surface and ω is the Coriolis parameter, included in the manner indicated by Dronkers.[2]

Inherent in Equations 5.5 and 5.6 are the assumptions that the pressure variation within the fluid is hydrostatic and atmospheric on the fluid surface.

Similarly, vertical integration of the continuity equation together with the boundary conditions

$$\frac{\partial h}{\partial t} + u\frac{\partial h}{\partial x} + v\frac{\partial h}{\partial y} - w = 0 \qquad \text{at } z = h$$

and

$$u\frac{\partial H}{\partial x} + v\frac{\partial H}{\partial y} - w = 0 \qquad \text{at } z = -H$$

definitions (5.7) and (5.8), leads to,

$$\frac{\partial h}{\partial t} + \frac{\partial}{\partial x}[(H + h)U] + \frac{\partial}{\partial y}[(H + h)V] = 0 \tag{5.9}$$

Equations 5.5, 5.6 and 5.9 may then be solved subject to sufficient and necessary boundary conditions (Figure 5.2)

$$(h)_{B_1} = h_{B_1} \quad \text{(specified height)} \tag{5.10}$$

$$(Ul_x + Vl_y)_{B_2} = V_n^* \tag{5.11}$$

where l_x and l_y are direction cosines.

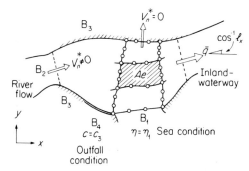

Figure 5.2 Typical domain and boundary conditions

Note that for the general case of the fully reflecting boundary, designated B_3 in Figure 5.2, the normal velocity $V_n^* = 0$ and

$$(Ul_x + Vl_y)_{B_3} = 0 \tag{5.12}$$

The initial condition defined throughout the domain under consideration is

$$h(x, y, t = 0) = h_0(x, y) \tag{5.13}$$

$$U(x, y, t = 0) = U_0(x, y) \tag{5.14}$$

$$V(x, y, t = 0) = V_0(x, y) \tag{5.15}$$

5.2.2 Convective-diffusion

The convective-diffusion equation for a time averaged (i.e. ignores turbulence based fluctuations) distribution of the concentration of effluent can be written, $i = 1, 2, 3 \cdots ll$,

$$\frac{\partial \bar{c}^i}{\partial t} + \frac{\partial}{\partial x}(\bar{c}^i U) + \frac{\partial}{\partial x}(\bar{c}^i V) - \frac{\partial}{\partial x}\left(e_x \frac{\partial \bar{c}^i}{\partial x}\right) + \frac{\partial}{\partial y}\left(e_y \frac{\partial \bar{c}^i}{\partial y}\right) + \frac{\partial}{\partial z}\left(e_z \frac{\partial \bar{c}^i}{\partial z}\right)$$

$$- q^i + S^i \bar{c}^i = 0 \tag{5.16}$$

where $\bar{c}^i$ refers to the concentration of the ith species, ll is the total number of species present, q is a source term and s a dissipation factor per unit volume and e_x, e_y, e_z denote the eddy diffusion coefficient in the appropriate coordinate axis direction.

Assuming a *well mixed* estuary, $e_z = 0$, and vertical integration is valid, we may define,

$$c^i = \frac{1}{(H+h)} \int_{-H}^{h} \bar{c}^i \, dz = \frac{1}{(H+h)} \langle \bar{c}^i \rangle \qquad (5.17a)$$

$$c^i = c^i(1 + \xi(z)) \qquad (5.17b)$$

where $\int_{-H}^{h} \xi(z) \, dz = 0$ etc., such that Equation 5.16 becomes,

$$\frac{\partial}{\partial t}((H+h) \cdot c^i) + \frac{\partial}{\partial x}\{\langle 1 + \beta_1 \xi \rangle \cdot U c^i\} + \frac{\partial}{\partial y}\{\langle 1 + \beta_2 \xi \rangle V c^i\} - (H+h)Q$$

$$-(H+h) \cdot S^i c^i = \frac{\partial}{\partial x}\left\{(H+h) \cdot D_x^i \cdot \frac{\partial c^i}{\partial x}\right\}$$

$$+ \frac{\partial}{\partial y}\left\{(H+h)D_y^i \cdot \frac{\partial c^i}{\partial y}\right\} \qquad (5.18)$$

where Q, S and D are defined in a similar manner to above and D is now called a *dispersion* coefficient. Equation 5.18 can be rewritten by using the transformation

$$(H+h) = \langle 1 + \beta_1 \xi \rangle = \langle 1 + \beta_2 \xi \rangle = H_1$$

as

$$\frac{\partial}{\partial t}(H_1 c^i) + \frac{\partial}{\partial x}\{H_1 U c^i\} + \frac{\partial}{\partial y}\{H_1 V c^i\} - H_1 Q^i - H_1 S^i c^i$$

$$= \frac{\partial}{\partial x}\left\{H_1 D_x^i \frac{\partial c^i}{\partial x}\right\} + \frac{\partial}{\partial y}\left\{H_1 D_y^i \cdot \frac{\partial c^i}{\partial y}\right\} \qquad (5.19)$$

which contains the assumption that the distribution of β_1, β_2 and ξ do not vary spatially.

Expanding Equation 5.19 and combining with the continuity equation,

$$H \frac{\partial c^i}{\partial t} + H_1 U \frac{\partial c^i}{\partial x} + H_1 V \frac{\partial c^i}{\partial y}$$

$$-R^i = \frac{\partial}{\partial x}\left\{H_1 D_x^i \frac{\partial c^i}{\partial x}\right\} + \frac{\partial}{\partial y}\left\{H_1 D_y^i \frac{\partial c^i}{\partial y}\right\} \qquad (5.20)$$

where $R^i = H_1 Q^i - H_1 S^i c^i$.

Rewriting we have,

$$\frac{\partial c^i}{\partial t} + U\frac{\partial c^i}{\partial x} + V\frac{\partial c^i}{\partial y} - Q^i - S^i c^i$$

$$= \frac{1}{H_1}\left\{ \frac{\partial}{\partial x}\left(H_1 D^i_x \frac{\partial c^i}{\partial x}\right) + \frac{\partial}{\partial y}\left(H_1 D^i_y \frac{\partial c^i}{\partial y}\right) \right\} \qquad (5.21)$$

Subject to boundary conditions of the form (Figure 5.2)

$$\text{(i)} \quad c^i = c^i_4 \text{ specified on part boundary } B_4 \qquad (5.22a)$$

and

$$\text{(ii)} \quad D^i_x \cdot \frac{\partial c^i}{\partial x}l_x + D^i_y\frac{\partial c^i}{\partial y}l_y + \bar{q} = 0 \qquad (5.22b)$$

On part boundary B_2, where D^i_x and D^i_y are dispersion coefficients associated with the x and y directions, l_x and l_y are as defined previously and $\bar{q}$ represents the outward flux per unit of surface.

The initial condition takes the form

$$c^i(x, y, t = 0) = c^i_0(x, y) \qquad (5.22c)$$

Equations 5.5, 5.6, 5.9, 5.21 and 5.22, for the simplified integrated case, comprise the necessary and sufficient equations for setting up the required numerical model.

5.3 The formulation process

5.3.1 The method of weighted residuals

Using the Galerkin weighted residual process, as outlined in previous chapters, for a domain discretized into a number (n_e) of elements, the global assembly of elemental contributions associated with a differential operator $\mathcal{D}$ is,

$$\sum_{i=1}^{n_e} \iint_\Omega N_i\{\mathcal{D}(Q_a)\}\, d\Omega^e \qquad i = 1, 2, \ldots, n \qquad (5.23)$$

Ω^e is the associated area of an element e and N_i are appropriate shape functions.

Note, for convenience of symmetry the system (5.23) includes all boundary equations.

In order to develop a similar form to Equation 5.23 for the coupled tidal/dispersion problem it is necessary to define general spatial approximations to all the appropriate variables U, V, h, H, D^i_x, D^i_y, c^i, Q^i, S^i and ω

in the form

$$U_a = \mathbf{N}^T\mathbf{U} \tag{5.24}$$

where the summation is over a typical element and where, for convenience, but not necessarily, all shape functions are identical.

Substitution of the approximate solutions (5.24) into Equations 5.5, 5.6 and 5.21 yields,

$$\left.\begin{array}{l}\sum_{i=1}^{n_e}\iint_{\Omega^e} N_i\left[\dfrac{\partial U_a}{\partial t} + U_a\dfrac{\partial U_a}{\partial x} + V_a\dfrac{\partial U_a}{\partial y} - \omega_a V_a + g\dfrac{\partial}{\partial x}(h_a - h_a^1)\right.\\[3mm] \left.+g\dfrac{(U_a^2 + V_a^2)^{\frac{1}{2}}}{C_C^2(H_a + h_a)} - \dfrac{\tau_{xa}}{(H_a + h_a)}\right]d\Omega^e = 0\end{array}\right\} \tag{5.25a}$$

$$\left.\begin{array}{l}\sum_{i=1}^{n_e}\iint_{\Omega^e} N_i\left[\dfrac{\partial V_a}{\partial t} + U_a\dfrac{\partial V_a}{\partial x} + V_a\dfrac{\partial V_a}{\partial y} + \omega_a U_a + g\dfrac{\partial}{\partial y}(h_a - h_a^1)\right.\\[3mm] \left.+g\dfrac{(U_a^2 + V_a^2)^{\frac{1}{2}}}{C_C^2(H_a + h_a)} - \dfrac{\tau_{ya}}{(H_a + h_a)}\right]d\Omega^e = 0\end{array}\right\} \tag{5.25b}$$

$$\sum_{i=1}^{n_e}\iint_{\Omega^e} N_i\left[\dfrac{\partial h_a}{\partial t} + \dfrac{\partial}{\partial x}[(H_a + h_a)U_a] + \dfrac{\partial}{\partial y}[(H_a + h_a)V_a]\right]d\Omega^e \tag{5.25c}$$

$$\left.\begin{array}{l}\sum_{i=1}^{n_e}\iint_{\Omega^e} N_i\left[\dfrac{\partial c_a^i}{\partial t} + U_a\dfrac{\partial c_a^i}{\partial x} + V_a\dfrac{\partial c_a^i}{\partial y} - Q_a^i - S_a^i c_a^i\right.\\[3mm] \left.-\dfrac{1}{H_{1a}}\left(\dfrac{\partial}{\partial x}\left(H_{1a}D_{xa}\dfrac{\partial c_a^i}{\partial x}\right) + \dfrac{\partial}{\partial y}\left(H_{1a}D_{ya}^i\dfrac{\partial c_a^i}{\partial y}\right)\right)\right]d\Omega^e = 0\end{array}\right\} \tag{5.25d}$$

The second order differentials in Equation 5.25 impose unnecessary constraints on interelement compatibility and are therefore reduced, by the application of Green's theorem, which results in,

$$\left.\begin{array}{l}\sum_{i=1}^{n_e}\iint_{\Omega^e}\left\{N_i\left[\dfrac{\partial c_a^i}{\partial t} + U_a\dfrac{\partial c_a^i}{\partial x} + V_a\dfrac{\partial c_a^i}{\partial y} - Q_a^i - S_a^i c_a^i\right]\right.\\[3mm] +\left[\dfrac{\partial N_i}{\partial x}\cdot D_{xa}^i\dfrac{\partial c_a^i}{\partial x} + \dfrac{N_i}{H_{1a}}\cdot D_{xa}^i\cdot\dfrac{\partial H_{1a}}{\partial x}\cdot\dfrac{\partial c_a^i}{\partial x} + \dfrac{\partial N_i}{\partial y}\cdot D_{ya}^i\dfrac{\partial c_a^i}{\partial y}\right.\\[3mm] \left.\left.+\dfrac{N_i}{H_a}\cdot D_{ya}^i\dfrac{\partial H_a}{\partial y}\right]\right\}d\Omega^e - \int_\Gamma N_i\left(D_{xa}^i l_x\dfrac{\partial c_a^i}{\partial x} + D_{ya}^i l_y\dfrac{\partial c_a^i}{\partial y}\right)d\Gamma\end{array}\right\} \tag{5.25e}$$

$$i = 1, 2, 3\ldots n$$

The resulting matrices are assembled element by element.

The globally assembled equation system, in matrix form, is for any time level τ,

$$\mathbf{C}\cdot\dot{\mathbf{U}}_\tau + \mathbf{F}^l(U, V, h)_\tau = 0 \tag{5.26a}$$

$$\mathbf{C} \cdot \dot{\mathbf{V}}_\tau + \mathbf{F}^{II}(U, V, h)_\tau = 0 \qquad (5.26b)$$

$$\mathbf{C} \cdot \dot{\mathbf{h}}_\tau + \mathbf{F}^{III}(U, V, h)_\tau = 0 \qquad (5.26c)$$

and

$$\mathbf{C} \cdot \dot{\mathbf{c}}^i_\tau + \mathbf{F}^{IV}(U, V, h)_\tau = 0 \qquad (5.26d)$$

where $\mathbf{C}$ is a square symmetric banded matrix and the dot signifies a partial time derivative. A typical contribution to $\mathbf{C}$ is,

$$\mathbf{C}_{ij} = \sum_{p=1}^{n_e} \iint_{\Omega^e} N_i N_j \, d\Omega^e \qquad \begin{array}{l} i = 1, 2 \ldots n \\ j = 1, 2 \ldots n \end{array} \qquad (5.27)$$

and a contribution to $\mathbf{F}$ can be written

$$f_i^{III} = \sum_{p=1}^{n_e} \iint_{\Omega^e} \left\{ \left[\sum_{j=1}^{n} \frac{\partial N_j}{\partial x} \cdot H_{ij} + \sum_{j=1}^{n} \frac{\partial N_j}{\partial x} \cdot h_j \right] \sum_{k=1}^{n} N_k U_k \right.$$

$$+ \sum_{j=1}^{n} \left[\frac{\partial N_j}{\partial y} \cdot H_j + \frac{\partial N_j}{\partial y} \cdot h_j \right] \cdot \sum_{k=1}^{n} N_k V_k + \left[\sum_{j=1}^{n} (N_j H_j + N_j h_j) \right]$$

$$\times \left. \left[\sum_{k=1}^{n} \frac{\partial N_k}{\partial x} \cdot U_k + \frac{\partial N_k}{\partial y} \cdot V_k \right] \right\} d\Omega^e \qquad i = 1, 2 \ldots n \qquad (5.28)$$

In Equations 5.26 the column vectors $\mathbf{F}$ contain all the terms other than those associated with the time derivative.

Although computer storage requirements increase, it is computationally more convenient to assemble equations in a coupled form,

$$\mathbf{C}^* \dot{\mathbf{q}} \tau + \mathbf{F}^*(U, V, h, c)_\tau = 0 \qquad (5.29)$$

where the ith location in the q vector represents four equations in four unknowns,

$$q_i = \left\{ \begin{array}{c} \dot{U}_i \\ \dot{V}_i \\ \dot{h}_i \\ \dot{c}_i \end{array} \right\}$$

and $\mathbf{C}^*$ is now a banded 4×4 matrix where m is the total number of nodes in the domain. It is apparent, therefore, that all equations, including these relating to boundary nodes have been included in $\mathbf{C}^*$. This makes the equation easier to handle and also gives rise to matrix symmetry. Equation 5.29 now forms the basis for a suitable recurrence relationship if the initial conditions are known.

5.4 Time integration

In this chapter the authors limit the discussion to three types of time integration schemes as applied to the particular problem under investigation:

(a) The Adams–Moulton multi-step predictor-corrector procedure.
(b) The finite difference trapezoidal integration.
(c) The finite element in time.

5.4.1 The predictor-corrector method

Predictor-corrector methods provide useful algorithms for the numerical solution to the type of phenomenon represented by Equation 5.29, because of the relatively small number of derivative evaluations required. Stability is also an essential factor in the choice of a particular predictor-corrector process and algorithms with improved stability characteristics have been investigated.[11] One of these is the Adams–Moulton type with the condition that the corrector is iterated to approximate convergence. The predictor formula, given in terms of $\mathbf{q}$ is

$$\mathbf{q}_{t+4\Delta t} = \mathbf{q}_{t+3\Delta t} + \frac{\Delta t}{24}\{55\dot{\mathbf{q}}_{t+3\Delta t} - 59\dot{\mathbf{q}}_{t+2\Delta t} + 37\mathbf{q}_{t+\Delta t} - 9\dot{\mathbf{q}}_t\} \quad (5.30)$$

and the corrector formula

$$\mathbf{q}_{t+4\Delta t} = \mathbf{q}_{t+3\Delta t} + \frac{\Delta t}{24}\{9\mathbf{q}_{t+4\Delta t} + 19\mathbf{q}_{t+3\Delta t} - 5\dot{\mathbf{q}}_{t+2\Delta t} + \dot{\mathbf{q}}_{t+\Delta t}\} \quad (5.31)$$

where the subscripts indicate the time level relative to the reference time t, and Δt is a time interval.

The process involves writing the system (5.29) in the form

$$\dot{\mathbf{q}} = \mathbf{f}(U, V, h, c, t) \quad (5.32)$$

and employing the Runge–Kutta, fourth degree equations to provide all the necessary derivatives for formula (5.30). These may be written as,[12]

$$\mathbf{q}_{t+\Delta t} = \mathbf{q}_t + \frac{\Delta t}{6}[\mathbf{K}_0 + 2\mathbf{K}_1 + 2\mathbf{K}_2 + \mathbf{K}_3] \quad (5.33a)$$

where,

$$\mathbf{K}_0 = \mathbf{f}(t, U_t, h_t, V_t, c_t) \quad (5.33b)$$

$$\mathbf{K}_1 = \mathbf{f}\left(t + \frac{\Delta t}{2}, U_t + \frac{\Delta t}{2}\mathbf{K}_0, V_t + \frac{\Delta t}{2}\mathbf{K}_0, h_t + \frac{\Delta t}{2}\mathbf{K}_0\right) \quad (5.33c)$$

$$\mathbf{K}_2 = \mathbf{f}\left(t + \frac{\Delta t}{2}, U_t + \frac{\Delta t}{2}\mathbf{K}_1, V_t + \frac{\Delta t}{2}\mathbf{K}_1, h_t + \frac{\Delta t}{2}\mathbf{K}_1\right) \quad (5.33d)$$

$$\mathbf{K}_3 = \mathbf{f}(t + \Delta t, U_t + \Delta t\,\mathbf{K}_2, V_t + \Delta t\,\mathbf{K}_2, h_t + \Delta t\,\mathbf{K}_2) \quad (5.33e)$$

For all succeeding values of **q** in time the predictor is used only once. The updated derivative $(t + 4\Delta t)$ required for the corrector is found using Equation 5.32, and the corrector value $q_{t+4\Delta t}$ evaluated. If this is sufficiently close to the predicted value, the process continues by predicting an updated value $q_{t+5\Delta t}$. If however the convergence criterion is not satisfied then a further solution is obtained to Equation 5.32 using the unconverged corrector value to form the right-hand side. The corrector formula is again used and this latter process repeated until the required convergence tolerance is satisfied.

An indication of the per step, truncation error may be obtained from the expression[13]

$$-\tfrac{1}{14}(\dot{\mathbf{q}}_{t+4\Delta t} - \mathbf{q}_{t+4\Delta t}) \qquad (5.34)$$

so that the time interval may be adjusted, within the stability criterion, to allow for the greatest possible time step.

5.4.2 The finite difference trapezoidal rule

The system represented by Equation 5.29 may be integrated in time by replacing the time derivative by a truncated Taylor series. This effectively implies that the value q varies linearly within a typical time interval Δt. If it is further assumed that the function $F^*(u, v, h)$ also varies linearly then by defining the suffix τ at $t + \Delta t/2$

$$\mathbf{C}^*\dot{\mathbf{q}}_{t+\Delta t/2} = -\mathbf{F}^*(U, V, h, c)_{t+\Delta t/2} \qquad (5.35)$$

or

$$\mathbf{C}^*\mathbf{q}_{t+\Delta t} = \mathbf{C}^*\mathbf{q}_t - \frac{\Delta t}{2}\mathbf{F}^*(U, V, h, c)_{t+\Delta t} + \mathbf{F}^*(U, V, h)_t \qquad (5.36)$$

This is known as the trapezoidal rule and a solution depends on the convergence of an iterative procedure for $F^*(u, v, h)_{t+\Delta t}$.

5.4.3 The finite element in time[14,15]

The integral equations in time are derived using the weighted residual method of Galerkin. As before, but now in the time dimension, a variation is defined typically as,

$$q_a(t) = \sum_{i=1}^{n} A_i(t)q_i \qquad (5.37)$$

where n is the number of unknown coefficients associated with the subdomain.

To complete the definition, the time variation of the vector $\mathbf{F}^*$ within a time domain is

$$F_a^*(t) = \sum_{i=1}^{n} A_{i(t)}F_i^* \qquad (5.38)$$

Provided the interpolation polynomials $A_{i(t)}$ are known, Equations 5.37 and 5.38 are substituted into Equation 5.29 weighted with the functions $A_{i(t)}$ and integrated over a typical interval $t_k, t_k + \Delta t$ represented by time element k. Hence, for the complete time domain, without any *a priori* assumption on the system at time $t_k + \Delta t$,

$$\sum_{k=1}^{n_{te}} \int_{t_k}^{t+\Delta t} A_i \{ \mathbf{C}^* \dot{\mathbf{q}}_a + \mathbf{F}_a^*(U, V, h, c) \} \, dt = 0 \qquad i = 1, 2, \dots n \qquad (5.39a)$$

The initial and boundary conditions are initially indistinguishable and are indicated generally as

$$\mathbf{q} = \mathbf{q}_0, \qquad \dot{\mathbf{q}} = \dot{\mathbf{q}}_0 \qquad (5.39b)$$

$$\mathbf{F} = \mathbf{F}_0, \qquad \dot{\mathbf{F}} = \dot{\mathbf{F}}_0 \qquad (5.39c)$$

and n_{te} is the total number of time elements.

5.4.3.1 Linear Lagrange type interpolation Typically (Figure 5.3), for one element, in time with two nodes numbered 1 and 2 for clarity, representing

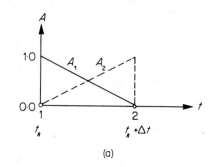

(a)

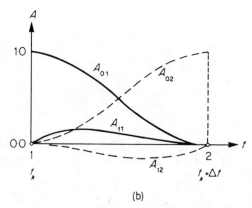

(b)

Figure 5.3 Time interpolation. (a) Linear time element. (b) Hermitian cubic functions

t_k and $t_k + \Delta t$,

$$q_{a(t)} = [A_1, A_2] \begin{Bmatrix} q_1 \\ q_2 \end{Bmatrix} \tag{5.40}$$

where

$$A_1 = 1 - \frac{(t - t_k)}{\Delta t} \tag{5.41a}$$

$$A_2 = \frac{(t - t_k)}{\Delta t} \tag{5.41b}$$

and

$$\dot{q}_{a(t)} = [\dot{A}_1, \dot{A}_2] \begin{Bmatrix} q_1 \\ q_2 \end{Bmatrix} \tag{5.42}$$

Substitution of Equations 5.40 through Equation 5.42 into Equation 5.39a gives

$$\sum_{k=1}^{n_{te}} \int_{t_k}^{t_k + \Delta t} A_i \mathbf{C}^k [\dot{A}_1, \dot{A}_2] \begin{Bmatrix} q_1 \\ q_2 \end{Bmatrix} + [A_1, A_2] \begin{Bmatrix} \mathbf{F}_1^k \\ \mathbf{F}_2^k \end{Bmatrix} dt = 0 \qquad i = 1, 2, \dots n \tag{5.43}$$

Equation 5.43 represents a complete set of assembled equations, valid for the whole time history of the problem. From an economic consideration only, the system is reduced to a single time element and the resulting equation solved in a step by step manner.

Equation 5.43 can now be written typically for the interval t_k, $t_k + \Delta t$, as

$$\int_{t_k}^{t_k + \Delta t} \begin{bmatrix} A_1 \mathbf{C} \dot{A}_1, & A_1 \mathbf{C} \dot{A}_2 \\ A_2 \mathbf{C} \dot{A}_1, & A_2 \mathbf{C} \dot{A}_2 \end{bmatrix} \begin{Bmatrix} q_1 \\ q_2 \end{Bmatrix} + \begin{Bmatrix} A_1 \\ A_2 \end{Bmatrix} [A_1, A_2] \begin{Bmatrix} F_1 \\ F_2 \end{Bmatrix} dt$$

$$k = 1, 2, \dots n_{te} \tag{5.44}$$

where the '*' notation has been omitted for clarity.

The initial value q_0 is known, which means that the variation of the principle (5.39a) with respect to this variable is identically zero, so that Equation 5.44 reduces to

$$\int_{t_k}^{t_k + \Delta t} [A_2 \mathbf{C} A_1, A_2 \mathbf{C} A_2] \begin{Bmatrix} q_1 \\ q_2 \end{Bmatrix} + [A_2 A_1, A_2 A_2] \begin{Bmatrix} F_1 \\ F_2 \end{Bmatrix} dt = 0 \tag{5.45}$$

Substitution from Equations 5.41a, 5.41b and 5.42 and integration over the interval, leads to the general recurrence relationship

$$\mathbf{C}^* q_{t_k + \Delta t} = \mathbf{C}^* q_{t_k} - \frac{\Delta t}{3} \mathbf{F}_{t_k}^* - \tfrac{2}{3} \cdot \Delta t \mathbf{F}_{t_k + \Delta t}^* \tag{5.46}$$

with the original notation.

For a time element with two nodes a general recurrence relationship is evolved of the form,

$$\mathbf{B}\hat{\mathbf{q}}_{t_k + \Delta t} = \mathbf{D}\{\hat{\mathbf{q}}_{t_k}, \hat{\mathbf{F}}^*_{t_k}, \hat{\mathbf{F}}^*_{t_k + \Delta t}\} \tag{5.47}$$

where '^' notation includes the time derivative, and the matrix $\mathbf{B}$ is unsymmetric. To solve for $\hat{\mathbf{q}}_{t_k + \Delta t}$ it is necessary to iterate for the updated values $\hat{\mathbf{F}}^*_{t_k + \Delta t}$ at each step. The interested reader is referred to the references cited above for further information relating to the formulation of coefficients in Equation 5.47.

Both methods outlined in this section rely on an iterative scheme in which convergence is accelerated by the relaxation method. An initial guess is successively corrected by a process of predicting a new solution q^{j+1} as a weighted function of the previous iterates q^j and q^{j-1}, so that,

$$\mathbf{q}^{j+1} = \mathbf{q}^{j-1} + \bar{\omega}(\mathbf{q}^j - \mathbf{q}^{j-1}) \tag{5.48}$$

and resolving the system of Equations 5.36, 5.46 and 5.47. Convergence was achieved when the difference between successive iterates was sufficiently small.

In the systems 5.36, 5.46 and 5.47 the matrices $\mathbf{C}^*$ and $\mathbf{B}$ are assembled and triangularized once only. This means that resolution involves forming an updated right-hand side, modifying this vector to suit the original triangularized system and back-substituting to complete the solution.

For the trapezoidal rule and linear finite element in time, empirical tests showed that $\bar{\omega} = 0.9$ provided an optimum convergence rate, whilst the cubic finite element in time algorithm converged most efficiently at $\bar{\omega} = 1.2$.

5.5 Stability and convergence

The analytical approach to the study of stability provides a guide to the performance of the time integration scheme used. Following a similar approach to methods given in References 2, 11, 12 the properties of 5.4.1, 5.4.2 and 5.4.3 can be conveniently summarized.[17]

(1) The stability of the multi-step predictor corrector sequence is determined by the corrector equation alone. The algorithm is conditionally stable with respect to the time interval.
(2) The trapezoidal integration scheme is unconditionally stable and no phase difference or damping is introduced by the method.
(3) The finite element in time is unconditionally stable but introduces spurious damping and phase retardation. This is more apparent in the linear finite element scheme.

5.6   Numerical examples

In order to assess the accuracy and practicability of the various algorithms discussed in Section 5.4, five examples will be considered. The first two are necessarily simple, chosen because they simulate tidal motion and have well defined mathematical solutions. Wind induced oscillation in a rotating lake is followed by the fourth example, tidal propagation in the southern North Sea, which illustrates fully two-dimensional flow. The final example comprises a typical estuary subjected to a continuous point effluent discharge.

Example 1

The first example compares the performance of the trapezoidal time integration scheme with the linear finite element in time in representing the propagation of an undamped sine wave in a rectangular channel 90 m × 15 m (Figure 5.4) and is not intended to represent a realistic long wave problem.

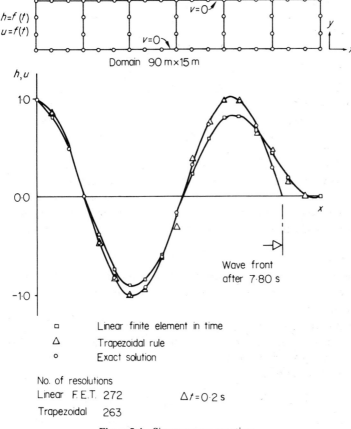

Figure 5.4   Sine wave propagation

Equations 5.5 and 5.6 are simplified to

$$\frac{\partial U}{\partial t} + g\frac{\partial h}{\partial x} = 0 \qquad (5.49)$$

$$\frac{\partial V}{\partial t} + g\frac{\partial h}{\partial y} = 0 \qquad (5.50)$$

$$\frac{\partial h}{\partial t} + H_1\frac{\partial U}{\partial x} + H_1\frac{\partial V}{\partial y} = 0 \qquad (5.51)$$

The boundary conditions are chosen so that no transverse motion occurs

$$h = A \sin (rt) \qquad\qquad x = 0, \quad 0 < y < 15, \quad t > 0$$

$$u = A \sin (rt) \qquad\qquad x = 0, \quad 0 < y < 15, \quad t > 0$$

$$V = 0 \qquad 0 < x < 90, \qquad y = 0, \quad 15, \quad t > 0$$

where $A = 1{\cdot}0$ m, $r = 1$ rad/s.
The solution to 5.49 and 5.51 is now

$$h, U = \sin (t - 0{\cdot}1x) \qquad (5.52)$$

provided the acceleration due to gravity is taken as $10\,\text{m/s}^2$ and the un-disturbed depth $H = 10$ m. The solution represents a sine wave propagating to $x = \infty$. These boundary conditions are chosen only because the solution is known *a priori* and they ensure that numerically u and h are exactly in phase. Other boundary conditions are considered in Example 2.

Six cubic isoparametric elements[17] represent the channel.

The results after $7{\cdot}8$ s are shown in Figure 5.4. The time interval is $0{\cdot}2$ s. As expected, the linear finite element in time damps the waveform whilst the trapezoidal rule preserves the oscillation.

Example 2

The second example is chosen to illustrate a comparison between the multi-step predictor-corrector sequence (R.K.P.C.) and the trapezoidal integration.

The equations of Example 1 are applicable and a rectangular channel 60 m $\times$ 15 m is chosen to represent a partly closed basin with one end ($x = 0$) open to the sea. Four parabolic isoparametric elements are used. The boundary conditions are

$$h = 1{\cdot}0(\sin (t)) \qquad\qquad x = 0, \quad 0 < y < 15, \quad t > 0$$

$$u = 0 \qquad\qquad\qquad x = 60, \quad 0 < y < 15, \quad t > 0$$

$$v = 0 \qquad 0 < x < 60, \qquad y = 0, \quad 15 \qquad\qquad t > 0$$

Tidal Propagation and Dispersion in Estuaries 111

These conditions represent a tidal variation at the open end and the normally assumed 'free-slip' velocity condition on the internal walls. The physical constants are $g = 9.81 \text{ m/s}^2$ and $H = 9.81$ m.

The solution is made up of a progressive wave and a reflective wave. Only the progressive wave will be considered here and can be shown to be

$$u, h = 1.0(\sin (t - x/9.81)) \qquad (5.53)$$

The results are shown in Figure 5.5. For profiles (a) $\Delta t = 0.3926$ s, and for profiles (b) $\Delta t = 0.5$ s.

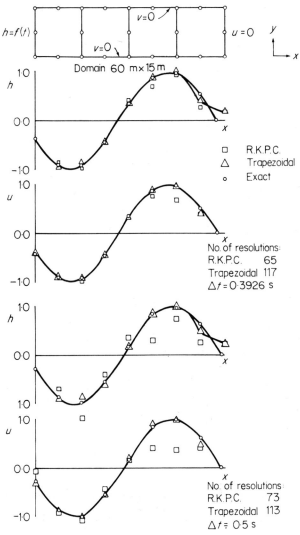

Figure 5.5 Sine wave propagation. (a) h, u profiles at 5·89 s; (b) h, u profiles at 6·00 s

The number of resolutions indicate the relative efficiency of the predictor-corrector process but accuracy is not maintained as the time interval is increased.

The conclusion that may be drawn from these simple experiments is that

(a) the linear finite element in time inherently damps the dominant waveform;
(b) the cubic finite element in time involves non-symmetric matrices and is too costly to be considered for realistic tidal propagation. The accuracy however is good;
(c) the multi-step predictor-corrector sequence is the cheapest algorithm on a per set basis but requires relatively smaller time steps for a given order of accuracy;
(d) the trapezoidal rule emerges well and will accurately predict a waveform both in amplitude and phase.

No mention has been made of the effect of dividing the domain into a greater or fewer number of elements. In fact for Example 2, four parabolic elements represent the coarsest mesh. Fewer elements fail to model the particular waveform with sufficient accuracy.[17]

Example 3

The trapezoidal rule was used to predict the wind induced mode of oscillation in a rectangular lake 125 km × 31·25 km. The lake is orientated in an east–west direction and the following properties apply

$$H = 80 \text{ m}$$

$$\tau_x = 0.0001 \text{ m}^2/\text{s}^2$$

$$\tau_y = 0$$

$$\omega = 0.0001/\text{s}$$

$$\Delta t = 100 \text{ s}$$

Equations 5.5, 5.6 and 5.9 were used, together with the boundary condition

$$u = 0 \quad x = 0, \quad 125 \text{ km}, \quad 0 \leqslant y \leqslant 31.25 \text{ km}, \quad t \geqslant 0$$

$$v = 0 \quad 0 \leqslant x \leqslant 125 \text{ km}, \quad y = 0, \quad 31.25 \text{ km}, \quad t \geqslant 0$$

This represents a free slip condition.

The initial condition for still water was

$$u = v = h = 0$$

Figure 5.6 shows longitudinal water surface profiles along the centre line of the lake at various times T in seconds after the application of the wind force at $T = 0$. The water surface oscillates about an equilibrium position

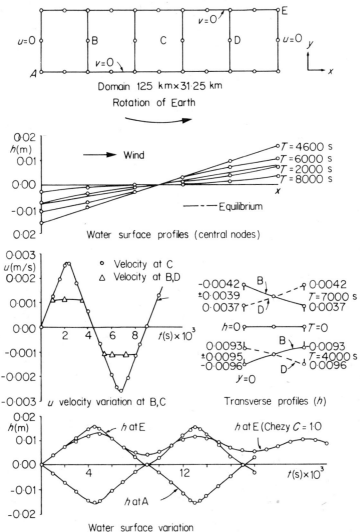

Figure 5.6 Wind induced oscillations in a rotating channel

where it finally comes to rest if bed friction had been included. This is indicated in the plot of surface elevation against time at point E when the Chézy coefficient $C_C = 1\cdot0$. This unrealistic value for C_C is used only to illustrate the effect of friction more clearly. The undamped period of oscillation of the lake is 8924·02 s and this is accurately predicted. The stream velocity (u) is shown to vary at half this period so that a maximum displacement occurs at zero streams.

The lateral velocity (v) is very small and arises from the Coriolis effect which also produces lateral variations in surface profile.

Example 4

The southern North Sea was chosen as a practical example of two-dimensional tidal flow including convective terms. The area was subdivided into 9 cubic isoparametric elements as shown in Figure 5.7.

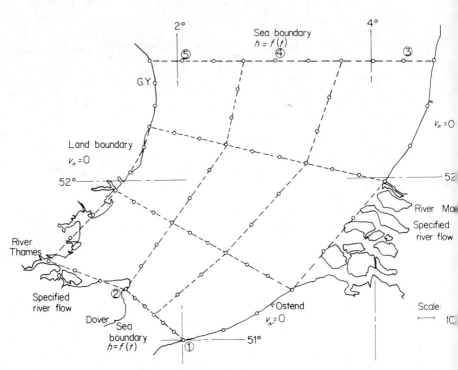

Figure 5.7 Domain and boundary conditions

Water levels, for September 12/13, 1958, at sea boundary points 1–5 were obtained from Reference 3. The levels at the remaining sea nodes were obtained by linear interpolation or extrapolation from this data. The resulting tidal data was expressed for each relevant node in the form of a polynomial using a least squares curve fitting routine. The land boundary assumed the no slip condition $u = v = 0$, $t \geqslant 0$. The discharge from the Thames and Maas/Rhine complex was considered constant and taken as the monthly average discharge over a period of 27 years.[18] For initial zero values everywhere within the domain, 6 cycles were necessary to obtain cyclic reproduction of results.

The following physical properties, specified as nodes, complete the data

Chézy coefficient $C_C = 19.4 \ln (0.9(H + h))$ (Reference 3)

$$\omega = 0.0001/s$$

$$\tau_x, \tau_y = 0$$

Depths H, averaged from *Admiralty Chart No. 2182A*, were specified at internal nodes, and a constant depth of 3 m simulated the coastal regions.

Computations were started with all velocities and water levels taken as zero.

Interpolated currents at nine points within each element are also shown.

Scale
→ 1 m/s

Figure 5.8 North Sea currents at 16·00 hours (interpolated currents at nine points within each element are also shown)

The problem was run using the trapezoidal integration scheme with a time interval of 6 minutes on the Atlas Computer Installation, Didcot, England. A calendar time of 24 hours was simulated in 96 minutes Central Processor time. This represents an average of 26 seconds per step.

A typical current plot is reproduced in Figure 5.8. Figure 5.9 shows the equal high tide times (co-tidal lines) and the co-range lines.

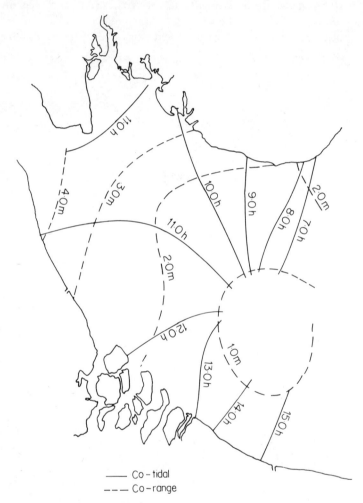

——— Co-tidal
--- Co-range

Figure 5.9 Co-tidal and co-range lines

The results agree essentially with those of Reference 3 and Figure 5.9 shows the amphidromic system with counterclockwise rotating tide associated with this part of the North Sea.

No attempt was made to adjust the model as detailed tidal information was lacking.

Example 5

A contour plot of the concentration distribution in a hypothetical, well-mixed estuary at 7·5 h after low water, is shown in Figure 5.10. The estuary is 15 000 metres (m) long and varies in width from 1000 m to 4000 m at the

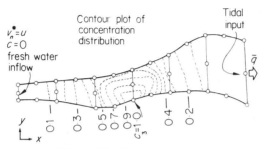

Figure 5.10 Hypothetical estuary

mouth. The M_2 tidal component, $\eta_1 = \sin(1{\cdot}405 \times 10^{-4} \times t)$ is used to simulate conditions at sea and a fresh water discharge of 2000 m³/s provides not only the inland boundary condition but also the initial velocity distribution. The initial water displacement $\eta(x, y, 0) = 0$ and the bed depth varies from 20 m inland to 80 m at the mouth. The Chézy friction coefficient $C_C = 50 \text{ m}^{\frac{1}{2}}/\text{s}$. A concentration denoted by $c_3 = 1{\cdot}0$ represents a continuous conservative waste discharge. The dispersion coefficients are $D_x = 10\,000 \text{ m}^2/\text{s}$ and $D_y = 800 \text{ m}^2/\text{s}$.

5.7 Conclusions

It has been demonstrated that the finite element method can be used, with advantage, to reproduce physical processes of tidal motion and the diffusion of effluents. The resulting ecosystem can be described in matrix form and standard techniques employed to evolve a recurrence relationship. During the present application a single matrix included all the pertinent equations. However, if desirable, the tidal equations could have been solved first and the resulting information utilized in the dispersion equation(s).

When solving the particular problems included in the text no difficulties were encountered regarding stability and convergence. However, for more complex systems it seems apparent that round-off can play a major part in the accuracy of the results. This is not surprising when the complexity of large problems necessitates a large number of iterations—within and per time step. Such errors can usually be circumvented by solving for the tidal and dispersion equations separately.

References

1. J. J. Dronkers, *Tidal Computations*, North Holland Publishing Company, Amsterdam, 1964.
2. J. J. Dronkers, 'Tidal computations for rivers, coastal areas and seas', *Journal Hydraulics Division, Proc. A.S.C.E.*, **95**, HY1, 29–77 (Jan. 1969).
3. J. J. Leendertse, 'Aspects of a computational model for long-period water wave propagation', *RAND Memorandum*, RM-5294-PR, Santa Monica, California (1967).
4. N. S. Heaps, ' A two-dimensional numerical sea model', *Philosophical Transactions of the Royal Society*, Series A, **265**, 93–137 (1969).
5. J. R. Rossiter and G. W. Lennon, 'Computation of tidal conditions in the Thames Estuary by the initial value method', *Proc. Institution of Civil Engineers*, **31**, 25–56 (1965).
6. J. R. Rossiter, 'Interaction between tide and surge in the Thames', *Geophysical Journal Royal Astr. Society*, **6**, 1, 29–53 (1961).
7. Lai Chintu, 'Numerical simulation of wave-crest movement in rivers and estuaries', *Proc. of the Tucson Symposium*, Vol. II, UNESCO, 699–713 (1969).
8. J. Ellis, 'Unsteady flow in channels of variable cross section', *Journal of the Hydraulics Division, Proc. A.S.C.E.*, HY10, 1927–1945 (Oct. 1970).
9. A. T. Ippen, 'Estuary and Coastline Hydrodynamics', McGraw-Hill, 1966.
10. B. Finlayson, *The Method of Weighted Residuals and Variational Principles*, Academic Press, New York, 1972.
11. R. R. Brown, J. D. Riley and M. B. Morris, 'Stability properties of Adams–Moulton type methods', *Math. of Comput.*, **19**, 90–96 (1965).
12. A. Ralston, *A First Course in Numerical Analysis*, McGraw-Hill, 1965.
13. G. L. Guymon, 'A finite element solution of the one-dimensional diffusion-convection equation', *Water Resources Research*, **6**, 1, 204–210 (1970).
14. J. H. Argyris and D. W. Scharpf, 'Finite elements in time and space', *The Aero. Journal of the Royal Aero. Society*, **73**, 1041–1044 (Dec. 1969).
15. I. Fried, 'Finite element analysis of time dependent phenomena', *A.I.A.A. Journal*, **7**, 6, 1170–1173 (1969).
16. I. Ergatoudis, B. M. Irons and O. C. Zienkiewicz, 'Curved isoparametric quadrilateral elements for finite element analysis', *Int. J. Solids and Structures*, **4**, 31–42 (1968).
17. J. M. Davis, Ph.D. thesis, University of Wales, 1974.
18. UNESCO, Discharge of Selected Rivers of the World, Vol. 1, 1969.

Chapter 6

Finite Element Lake Circulation and Thermal Analysis

R. H. Gallagher

6.1 Introduction

Industrial development is increasingly confronted with the demand for environmental impact studies prior to the approval of construction. Especially difficult problems are posed in the case of nuclear power plants with respect to the dissipation of heat, since high temperatures are critical to power plant efficiency. In the common case of a power plant that must be located adjacent to a freshwater lake, a number of alternative schemes for dissipation of the heat from the cooling system are possible, including cooling towers, cooling wells and the discharge of the heated effluent directly into the lake.[1,2]

The option of direct discharge of the effluent into the lake may have the least serious effects on the environment, taking into account the economic factors of the project, but even this solution will present many difficulties which must be considered in an environmental impact study. The objective of this chapter is to review the finite element procedures available for calculation of the temperature distributions due to the above circumstances.

The overall problem is a complicated one and it includes certain physical processes which are not yet fully understood. Also, finite element solutions of aspects of the problem which are of key importance have not yet been published. This is particularly true of the history of temperature in the immediate vicinity of the heated outfall. Nevertheless, the progress which has been made in finite element analysis has been rapid and it promises to move ahead still further in the immediate future.

The fundamental physical problem with which this review is concerned is the flow, as defined by the distribution of velocities. The steady state portion of this flow is caused by the action of the wind on the surface of the lake. Basic methods of finite element analysis of the wind-driven circulation of lakes are therefore first reviewed. This description, for the most part, parallels developments recently published by the writer and his colleagues;[3] a number of alternative finite element representations are also identified. This analysis technology has reached a point where it outdistances the field data available for its verification.

Various features of practical importance are neglected in the basic representations of wind-driven lake circulation. These include the presence of islands, the existence of thermal stratification and the three-dimensional character of the real lake. A section is devoted to examination of finite element developments which seek to account for these factors.

For special circumstances it is possible to employ the simplest finite element representations of heat transport in a body of water. The velocities are calculated by means of the flow analysis procedures cited above and are used in the discretized form of the heat transport equations. A brief outline is given of published results in this type of analysis. The chapter concludes with comments regarding the sources of information about the physical conditions at the outfalls of heated effluents.

6.2 Wind-induced lake circulation. Two-dimensional

In this section we first review the construction of a two-dimensional finite element representation of the steady state wind-driven circulation of shallow lakes and basins.

The cross-section shown in Figure 6.1 defines the basic geometric parameters of this development, which is due to Liggett and Hadjitheo-dourou[4] in its fundamental theoretical form. The origin of coordinates is

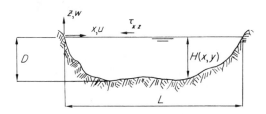

Figure 6.1

fixed at the surface of the lake, with z direction velocity $w = 0$ at $z = 0$. (The 'rigid lid' assumption.) The x and y direction velocities are u and v, respectively. The physical properties of the lake, including the eddy viscosity μ and the mass density per unit volume ρ, are assumed to be constant and the Coriolis parameter ω is also assumed constant. The pressure p is taken to vary hydrostatically. The surface wind stresses τ_{xz} and τ_{yz} are prescribed. Under these assumptions the direction momentum equations and the continuity equation take the form

$$-fv = -\frac{1}{\rho}\frac{\partial p}{\partial x} + \frac{\partial^2 u}{\partial z^2}\cdot\mu \qquad (6.1)$$

$$-fu = -\frac{1}{\rho}\frac{\partial p}{\partial y} + \frac{\partial^2 v}{\partial z^2}\cdot\mu \tag{6.2}$$

$$g = -\frac{1}{\rho}\frac{\partial p}{\partial z} \tag{6.3}$$

$$\frac{\partial u}{\partial x} + \frac{\partial v}{\partial y} + \frac{\partial w}{\partial z} = 0 \tag{6.4}$$

g is the acceleration due to gravity. A stream function ψ is defined as follows

$$\frac{\partial \psi}{\partial y} = UH, \qquad \frac{\partial \psi}{\partial x} = VH \tag{6.5}$$

in which U and V are depthwise averages of the component velocities. After combination of the above equations, with consideration of the boundary conditions (zero velocity on all solid surfaces and on the surfaces $\tau_{xz} = \mu\,\partial u/\partial x$, $\tau_{yz} = \mu\,\partial v/\partial y$) one obtains

$$\frac{\partial^2 \psi}{\partial x^2} + \frac{\partial^2 \psi}{\partial y^2} + A(x,y)\frac{\partial \psi}{\partial x} + B(x,y)\frac{\partial \psi}{\partial y} + C(x,y) = 0 \tag{6.6}$$

The terms $A(x,y)$, $B(x,y)$ and $C(x,y)$ are those which result from consideration of the varying depth and, as indicated, are functions of the planform coordinates.

Transformation of Equation 6.6 into the finite element form, given in detail in Reference 3, is accomplished by means of the method of weighted residuals with Galerkin criterion. The approximation of ψ is by means of the trial function $\bar{\psi}$, which has the form

$$\bar{\psi} = \sum N_i\psi_i = \mathbf{N}\boldsymbol{\psi} \tag{6.7}$$

wherein N_i are the shape functions and $\boldsymbol{\psi}$ are the nodal values of the stream function. Applying now the weighted residual concept

$$\int_A \mathbf{N}\left[\frac{\partial^2 \mathbf{N}}{\partial x^2} + \frac{\partial^2 \mathbf{N}}{\partial y^2} + A\frac{\partial \mathbf{N}}{\partial x} + B\frac{\partial \mathbf{N}}{\partial y} + C\right]\mathrm{d}A\{\boldsymbol{\psi}\} = 0 \tag{6.8}$$

Next, integration by parts in the plane is applied to reduce the order of the derivatives appearing in this integral and to produce boundary terms. One obtains

$$\int_A \left[\left(-\frac{\partial \mathbf{N}}{\partial x}\frac{\partial \mathbf{N}}{\partial x} - \frac{\partial \mathbf{N}}{\partial y}\frac{\partial \mathbf{N}}{\partial y} + A\mathbf{N}\frac{\partial \mathbf{N}}{\partial x} + B\mathbf{N}\frac{\partial \mathbf{N}}{\partial y}\right)\boldsymbol{\psi} + \mathbf{N}C\right]\mathrm{d}A$$

$$+ \oint \mathbf{N}\frac{\partial \mathbf{N}}{\partial n}\boldsymbol{\psi}\,\mathrm{d}S = 0 \tag{6.9}$$

The values of ψ are zero on the entire exterior boundary and the closure integrals along interelement boundaries vanish if continuity is preserved across these boundaries. Thus, the contour integral term is excluded from subsequent consideration. Evaluation of the remaining integrals for all i then yields the following system of element equations

$$\mathbf{k}^e \boldsymbol{\psi} = \mathbf{f}^e \tag{6.10}$$

in which

$$\mathbf{k}^e = \int_A \left(-\frac{\partial \mathbf{N}}{\partial x} \frac{\partial \mathbf{N}}{\partial x} - \frac{\partial \mathbf{N}}{\partial y} \frac{\partial \mathbf{N}}{\partial y} + A\mathbf{N} \frac{\partial \mathbf{N}}{\partial x} + B\mathbf{N} \frac{\partial \mathbf{N}}{\partial y} \right) dA \tag{6.11}$$

$$\mathbf{f}^e = -\int \mathbf{N} C \, dA \tag{6.12}$$

It should be noted that due to the terms $A\mathbf{N} \, \partial\mathbf{N}/\partial x$ and $B\mathbf{N} \, \partial\mathbf{N}/\partial y$ the resulting algebraic equations will be non-symmetric. Hence, advantage cannot be taken of symmetry as in finite element structural analysis. Also, as noted above, A, B and C are functions of y and x.

The equations for the complete lake are constructed from the equations of the elements by imposing the condition of stream function continuity at each element joint, which is synonymous with simple addition of all coefficients with like subscripts. Thus, the global equations are

$$\mathbf{K}\boldsymbol{\psi} = \mathbf{P} \tag{6.13}$$

After solution of this equation for ψ, the other variables, such as averaged velocities and pressure gradients, can subsequently be evaluated by back substitution.

Numerical solutions to Equation 6.13 have been obtained for both simple test problems and for Lake Ontario.[3] Since field data is not available for Lake Ontario the convergence of the solution has been studied with use of higher-order elements.[6] Cheng[9] has analysed Lake Erie, using a formulation which excludes consideration of variable depth. Tong[8] includes this factor in a finite element formulation based on Welander's theory,[7] which does not differ significantly from the theory cited above.

It is relevant to note the physical considerations involved in the action of the wind on the surface of the lake. Under ideal conditions, where the lake surface is calm and the wind direction and velocity (U_w) is invariant with time, the shear at the lake surface (τ_s) is given by

$$\tau_s = \frac{1}{2} C_f \rho_w U_w^2 \tag{6.14}$$

where ρ_w is the density of the air and C_f is the skin friction coefficient. From

Blasius's equation

$$C_f = 0.045 \left(\frac{v_w}{U_w d_w} \right) \qquad (6.15)$$

In this equation v_w is the kinematic viscosity of the air in ft^2/s and h_w is the thickness of the air boundary layer above the water surface. A correlation of extensive field data and analytical representations of wind action can be found in the recent paper by Wu.[5]

6.3 Extensions and special considerations in wind-induced lake circulation analysis

Of the numerous simplifications introduced in the analysis model described in the previous section, three take on particular importance: (1) the presence of islands, (2) the existence of thermal stratification during the summer season and (3) the three-dimensional character of the real body of water.

If a stream function is adopted as the independent variable, as is done in the formulations discussed previously, the presence of islands raises a basic complication in the definition of the boundary conditions at the node points of the island shore line. The stream function is zero at points on the shore of the lake but takes on a constant, undefined value on each of the islands. Thus, as Tong[8] proposes, the values of the stream function on a given island are set equal to a single value that is determined in the solution process. This substantially contracts the number of unknowns in the equations to be solved.

Cheng[9] adopts a different approach to the treatment of islands. The system of global equations is first assembled without consideration of the islands and their boundary conditions. We denote this solution as Ψ_0. Then, in succession, 'unit' solutions Ψ_i ($i = 1, \ldots M$, where M is the number of islands) are obtained for $\Psi_j = 1$ for node points on the respective islands. Finally, an $M \times M$ system of equations must be solved to give the amplitudes G_i ($i = 1, \ldots M$) which apply to the unit solutions. The complete solution is then represented by

$$\Psi = \Psi_0 + \sum_{i=1}^{M} G_i \Psi_i \qquad (6.16)$$

Thermal stratification is widely believed to exert an important influence on lake flow phenomena through its effects on density variations and other physical factors. In many lakes uniform temperature conditions are realized in winter and as summer atmospheric conditions approach a rise in temperature occurs in the upper regions of the lake. The peak is reached in these regions towards the end of summer. Since the rise in temperature penetrates to only a limited depth (say 20 to 40 feet) the lower portions of the lake are not affected, and a somewhat 'stratified' temperature profile prevails. The heated upper region is known as the epilimnion while the unheated lower region is termed the hypolimnion.

Basic to the analytical solution of this problem is the calculation of the above-described temperature profile. Although variations of this profile exist in the horizontal plane, such variations will generally be of no importance to the flow problem. Thus, the thermal analysis is one-dimensional and presents no computational difficulties. Many interpretations of available physical data for this purpose have been proposed, as summarized by Sundaram and Rehm,[10] and these authors, as well as Orlob and Selna,[11] Huber, Harleman and Ryan,[12] have conducted analyses of varying degrees of sophistication, generally on the basis of the finite difference approximation. Sundaram and Rehm have proposed an approach which describes a non-linear variation of the temperature (T) across the lake depth by means of the equation

$$\frac{\partial T}{\partial t} = \frac{\partial}{\partial z} e_z \frac{\partial T}{\partial z} \qquad (6.17)$$

where t is time and e_z is the applicable vertical eddy diffusivity. In this representation one of the relationships adopted for the latter is that due to Holzman[13]

$$e_z = e_{z0}\left(1 + cz^2\left(\frac{\partial T}{\partial z}\right)\right) \qquad (6.18)$$

wherein e_{z0} and c are constants dependent upon the circumstances of the environment being modelled.

Sundaram and Rehm have solved Equation 6.17 for the annual cycle of temperature of Lake Cayuga, New York, with good agreement with available field measurements. Finite element solutions of the same problem have been performed at Cornell University. The distinction between the finite difference and finite element approaches is negligible in the one-dimensional case, but this work serves to demonstrate that significant improvements in computational efficiency are possible and that the problem is treatable within the framework of a more general finite element computer program.

From the above it appears that the stratified temperature distribution is capable of prediction through analysis, although improvements in available data are desirable, and that the effects on physical properties entering into the flow equations can be defined. The state of affairs is less satisfactory in the flow analyses in which these properties find use. The relatively few flow solutions of the stratified lake problem, by Liggett and Lee,[14,15] Stuart[16] and Welander,[17] employ two-layered approximations, wherein the assumption is made that the epilimnion is of one constant density and the hypolimnion is of another density, and the two do not mix.

The two-layer lake is, of course, an idealization which is never realized in nature. The choice of layers largely dictates the flow patterns which result and it is important to establish the significance of this factor. Lee[18] has attempted to account for it by formulation and solution of the wind-driven circulation of a continuously stratified lake. A series solution, restricted to a rectangular basin, was accomplished.

It has already been emphasized that the general environmental problem will require that account be taken of the irregular topography of natural lakes, so that the need for a finite element solution for the continuously stratified lake is apparent. One step in this direction has been taken by Bedford[19] who has effected the solution for stratified circulation in a cross-section due to imposed surface shear stress and a predefined depthwise density variation. The formulation can be characterized as pertinent to the two-dimensional non-homogeneous cavity flow problem. The basic equations in two dimensions are

$$u\frac{\partial u}{\partial x} + v\frac{\partial u}{\partial y} = -\frac{1}{\rho}\frac{\partial p}{\partial x} + \frac{\partial}{\partial x}\left(\mu\frac{\partial u}{\partial x}\right) + \frac{\partial}{\partial y}\left(\mu\frac{\partial u}{\partial y}\right) \tag{6.19}$$

$$u\frac{\partial v}{\partial x} + v\frac{\partial v}{\partial y} = -\frac{1}{\rho}\frac{\partial p}{\partial y} + \frac{\partial}{\partial x}\left(\mu\frac{\partial v}{\partial x}\right) + \frac{\partial}{\partial y}\left(\mu\frac{\partial v}{\partial y}\right) \tag{6.20}$$

$$\frac{\partial u}{\partial x} + \frac{\partial v}{\partial y} = 0 \tag{6.21}$$

$$u\frac{\partial \rho}{\partial x} + v\frac{\partial \rho}{\partial y} = \frac{\partial}{\partial x}\left(e\frac{\partial \rho}{\partial x}\right) + \frac{\partial}{\partial y}\left(e\frac{\partial \rho}{\partial y}\right) \tag{6.22}$$

where μ is the eddy viscosity and e the eddy diffusivity and all other terms are as previously defined. The Boussinesq approximation has been invoked. Equation 6.22 permits density diffusion.

The above equations are operated upon to yield a contracted set of equations in terms of the stream function ψ. This equation, in dimensionless form, is

$$\frac{\mu}{Re}\nabla^4\psi + \frac{\partial.(\nabla^2\psi, \psi)}{\partial(x, y)} - \frac{Gr}{Re}\frac{\partial \rho}{\partial x} = 0 \tag{6.23}$$

and

$$\frac{-e}{Re\,Pr}\nabla^2\psi + \frac{\partial \psi}{\partial y}\frac{\partial \rho}{\partial x} - \frac{\partial \psi}{\partial x}\frac{\partial \rho}{\partial y} = 0 \tag{6.24}$$

in which *Re*, *Gr* and *Pr* are the Reynolds number, Grashof number and Prandtl number, respectively, defined as

$$Re = \frac{\tau D}{\rho_0\mu^2}, \qquad Gr = -\frac{\Delta\rho g D^3}{\rho\mu^2}, \qquad Pr = \frac{\mu}{e}$$

where D is the cavity depth, ρ_0 is a reference density, and $\Delta\rho$ is the density difference between top and bottom of the lake. The eddy viscosity is assumed constant ($\mu = $ const.). The flow is driven by a shear (τ) applied to the surface in the cavity and the no-slip condition is enforced on the solid boundaries. The diffusion of density is important to the phenomena being studied. In order to maintain a stratified field the density is held constant at the top and bottom surfaces and a density flux is maintained through the cavity. This condition is not unlike a stratified lake where a density flux is created by solar heating.

Discretization of this formulation in the finite element mode, using the method of weighted residuals with the Galerkin criterion, leads to an integral form involving second derivatives of the stream function. This poses the requirement of C^1 continuity, i.e. the chosen field of ψ should evidence continuity of both the function itself and its derivatives with respect to the normal directions of the element boundaries. Numerical computations to date have employed fields of the form of the displacement field chosen for familiar plate bending elements, where C^1 continuity requirements also prevail.

It should be noted that the difficulty of achieving C^1 continuity in plate bending elements has led to a variety of schemes with simpler representation requirements, including 'generalized' variational principles, 'discrete-Kirchhoff' formulations, and mixed and hybrid procedures. These appear to deserve consideration for the problem defined above.

The formulation of a finite element representation for the three-dimensional lake has been outlined by Leonard and Melfi.[20] They remove the 'rigid lid' assumption; the resulting formulation is then non-linear in the velocity w. The Newton–Raphson method, which requires the construction and inversion of a Hessian matrix consisting of second derivatives of the basic matrix equations, is proposed as the solution algorithm. Isoparametric elements in the form of hexahedra are proposed for the finite element idealization.

Reference 20 presents no numerical solutions. In the view of the writer it is doubtful whether three-dimensional representations of lake circulation, even in the linearized form, will be economically justified as components of more general studies of thermal effects. The value of such computations may reside in the analytical verification of the data accumulated in physical modelling of the lake circulation, such as those reported by Rumer and Hoopes.[21]

6.4 Transport of temperature or substances

The determination of the distribution by transport of temperature in a lake or basin with known flow is a problem of major practical importance. We have seen in the previous sections that wind-driven circulation is calculable under the assumptions stated therein. In the case of cooling ponds and similar basins the predominant flow is similarly calculable. Temperature distributions have been determined for such conditions by Loziuk, Anderson and Belytschko.[22,23] More recently, Tong[8] presented a more general development along these lines which permits the finite element calculation of any concentration of substance in a lake. We outline the latter in this section.

If we define ϕ as the average concentration across the depth (h) of the substance under study, the governing differential equation can be written as

$$\rho\left(\frac{\partial(h\phi)}{\partial t} + \frac{\bar{u}}{h}\frac{\partial(h\phi)}{\partial x} + \frac{\bar{v}}{h}\frac{\partial(h\phi)}{\partial y}\right) = \frac{\partial}{\partial x}D_x\frac{\partial(h\phi)}{\partial x} + \frac{\partial}{\partial y}D_y\frac{\partial(h\phi)}{\partial y} + Q \qquad (6.25)$$

where D_x and D_y are the turbulent diffusion coefficients and Q is a source or sink term. Now, the approximation of ϕ can be written in the form of the trial function

$$\bar{\phi} = \sum N_i\phi_i = \mathbf{N}\boldsymbol{\phi} \qquad (6.26)$$

where $\boldsymbol{\phi}$ represents nodal values of ϕ and $\mathbf{N}$ is the relevant set of shape functions. When the analysis is performed for temperature, with a single temperature (T) across the depth of the lake, $T = h\phi$.

Application of the weighted residual approach can again be made to construct algebraic equations for the element. When this is done, using Equations 6.25 and 6.26, one obtains

$$\mathbf{C}\dot{\boldsymbol{\phi}} + \mathbf{S}\boldsymbol{\phi} = \mathbf{Q} \qquad (6.27)$$

where

$$\mathbf{C} = \left[\int_A \rho\mathbf{N}\mathbf{N}\,\mathrm{d}A\right] \qquad (6.28)$$

$$\mathbf{S} = \left[\int_A \left[\mathbf{N}\left(\frac{\bar{u}}{h}\frac{\partial\mathbf{N}}{\partial x} + \frac{\bar{v}}{h}\frac{\partial\mathbf{N}}{\partial y}\right)\rho + k_x\frac{\partial\mathbf{N}}{\partial x}\frac{\partial\mathbf{N}}{\partial x} + k_y\frac{\partial\mathbf{N}}{\partial y}\frac{\partial\mathbf{N}}{\partial y}\right]\mathrm{d}A\right] \qquad (6.29)$$

The vector $\mathbf{Q}$ embodies the representation of the source or sink term and any of the pertinent prescribed boundary conditions. Finally, by assembly of the global equations from the element equations

$$\hat{\mathbf{C}}\dot{\boldsymbol{\phi}} + \hat{\mathbf{S}}\boldsymbol{\phi} = \hat{\mathbf{Q}} \qquad (6.30)$$

where $\hat{\mathbf{C}}$, $\hat{\mathbf{S}}$ and $\hat{\mathbf{Q}}$ correspond to $\mathbf{C}$, $\mathbf{S}$ and $\mathbf{Q}$.

The idealization for transport analysis is done in the same way as for flow analysis. After calculation of the velocities in the flow analysis the values obtained are used in the formation of the matrix $\mathbf{S}$.

Loziuk and coworkers[22,23] apply the above approach to various practical problems, including an actual lake with irregular boundary. Available field data indicate a reasonable level of agreement with the analysis results.

Tong[8] calculates the diffusion of a substance in a rectangular basin containing a circular island.

6.5 More general problems

Three separate analyses of the distribution of temperature due to heated water outlets appear to be necessary:

(a) the outlet region, or zone-of-flow-establishment region;
(b) the near field, or zone-of-established-flow;
(c) the far field.

The behaviour in the zone-of-flow-establishment is strongly influenced by the channel and shore geometries. In the near field the thermal plume has the characteristics of a fully developed jet; its behaviour is controlled by the initial momentum flux and imposed external effects. The initial momentum flux and other input aspects of this phase depend on the solution of the zone-of-flow-establishment phase. The far field is the total body of water and it is to this problem that the finite element method has been applied.

Analytical studies, by any method, of the zone-of-flow-establishment regime have been quite limited. Hirst,[27] by forming the integral equations of conservation of mass, energy and momentum, develops a set of five non-linear, coupled ordinary differential equations in the principal variables of the problem (e.g. centre line temperature, length of zone-of-flow establishment). Solutions of these five equations are obtained by standard numerical methods.

In contrast, a substantial number of studies have been made of the zone-of-established-flow. Stefan and Vaidayaraman[30] employ what is termed 'the entrainment principle' to form a set of three-dimensional equations which are solved numerically. Barry and Hoffman,[25] using finite differences, directly approximate the full set of three-dimensional governing equations, including the continuity equation and the equations of motion. Among the analytical solutions, Zeller, Hoopes and Rohlich[24] and Motz and Benedict[26] use integral relationships in two-dimensional representations.

In order to establish effective finite element representations for the outlet region and the near field it would be desirable to construct more concise governing relationships than those used in the above studies. Basic theoretical work and a careful study of physical data, rather than the usual approximation of a generally agreed-upon differential equation is therefore a part of finite element developmental efforts. Also, the procedure of forming and solving integral relationships, which is popular among the studies cited above, is not the route taken in conventional finite element analysis.

Physical modelling is presently viewed as the most appropriate method of obtaining data for environmental impact studies.[29] Furthermore, the data obtained from these studies indicate that satisfactory results are given by gross analytical models, such as those which deal in overall heat and energy balances.[28] Thus, more field measurements of physical data and of correlation with analytical procedures will be needed before it is known if the finite element method, or indeed any sophisticated numerical technique, will be of value for the full range of thermal pollution analyses. The physical circumstances may be too uncertain to justify the related computational expense.

6.6   Concluding remarks

A sufficient number of recent papers have been published to support the conclusion that the finite element method can be applied in a routine way to the solution of problems of lake circulation and to the one- and two-dimensional analysis of the transport of temperature and pollutants. If it is necessary to account for the varying depth of the lake it is possible to construct non-self-adjoint formulations in preference to the general three-dimensional formulations. In view of the uncertainty of physical data it is doubtful that more sophisticated representations are justified. A long period of correlation of numerical solutions obtained with available programs, with field measurements, is foreseen.

The more critical environmental problems associated with heated flow and the local dispersion of pollutants have not been tackled by the finite element method, at least in the already-published literature. Certain difficulties have been identified here, all of which are shared by alternative analytical approaches. An active area of research and application of the finite element method is therefore foreseen in this connection.

Acknowledgement

The review presented herein was prepared with support of the U.S. National Science Foundation under Research Grant GK-23992. The advice of the author's colleague, Prof. James A. Liggett, and of Drs. K. Bedford and S. T. K. Chan, is gratefully acknowledged.

References

1. G. Kinsman, 'Power plant cooling systems', *Proc. ASCE, J. Power Div.*, **98**, no. PO 2, 247–252 (1972).
2. F. K. Moore, 'Effects of a power plant on a deep stratified lake', *Bulletin of the American Physical Society* (Nov. 1970).
3. R. H. Gallagher, J. A. Liggett and S. T. K. Chan, 'Finite element circulation analysis of variable-depth shallow lakes', *Proc. ASCE, J. of the Hyd. Div.*, **99**, no. HY 7, 1083–1096 (1973).
4. J. A. Liggett and C. Hadjitheodorou, 'Circulation in shallow homogeneous lakes', *Proc. ASCE, J. of the Hyd. Div.*, **95**, no. HY 2, 609–620 (1969).

5. J. Wu, 'Prediction of near-surface drift currents from wind velocity', *Proc. ASCE, J. of the Hyd. Div.*, **99**, no. HY 9, 1291–1302 (1973).

6. R. H. Gallagher and S. T. K. Chan, 'Higher-order finite element analysis of lake circulation', *Computers and Fluids*, **1**, 1 (1973).

7. P. Welander, 'Wind action on shallow sea, some generalizations of Eckmann's theory', *Tellus*, **9**, 1, 47–52 (1957).

8. Pin Tong, 'Finite element solution of the wind-driven current and its mass transport in lakes', *Int. Conf. on Numerical Methods in Fluid Dynamics, University of Southampton, Sept. 1973.*

9. R. T. Cheng, 'Numerical investigation of lake circulation around islands by the finite element method', *Int. J. for Numerical Methods in Engrg.*, **5**, 1, 103–112 (1972).

10. T. Sundaram and R. Rehm, 'Formation and maintenance of thermoclines in stratified lakes, including the effects of power plant thermal discharges', *AIAA J.*, **9**, 7 (1971).

11. G. Orlob and L. Selna, 'Temperature variations in deep reservoirs', *Proc. ASCE, J. of the Hyd. Div.*, **96**, no. HY 2, 391–410 (1970).

12. W. Huber, D. Harleman and P. Ryan, 'Temperature prediction in stratified reservoirs', *Proc. ASCE, J. of the Hyd. Div.*, **98**, no. HY 4 (1972).

13. B. Holzman, 'The influence of stability on evaporation', *Proc. N.Y. Academy of Sciences*, **44**, p. 13 (1943).

14. J. A. Liggett and K. K. Lee, 'Properties of circulation in stratified lakes', *Proc. ASCE, J. of the Hyd. Div.*, **97**, no. HY 1 (1971).

15. K. K. Lee and J. A. Liggett, 'Computation of circulation in stratified lakes', *Proc. ASCE, J. of the Hyd. Div.*, **96**, no HY 10 (1970).

16. W. Stuart, 'Evaporate deposition in layered sea, a wind-driven dynamical model', Northwestern University Dept. of Engineering Sciences, Aug. 1971.

17. P. Welander, 'Wind-driven circulation in one- and two-layer oceans of variable depth', *Tellus*, **20** (1968).

18. K. K. Lee, 'Wind-induced and thermally-induced currents in the Great Lakes', *Water Resources Res.*, **8**, 6, 1444–1455 (1972).

19. K. Bedford, 'A numerical investigation of stably stratified, wind driven cavity flow by the finite element method', Ph.D. Dissertation, School of Civil and Environmental Engineering, Cornell University, 1974.

20. J. Leonard and D. Melfi, '3-D finite element model for lake circulation', *Proceedings of Third Conference on Matrix Methods in Structural Mechanics, Dayton, Ohio*, AFFDL TR 71-160, 1025–1048 (1973).

21. R. Rumer and J. Hoopes, *Modeling Great Lakes Circulations*, Report, Wisconsin Water Resources Center, 1971.

22. L. Loziuk, J. Anderson and T. Belytschko, 'Hydrothermal analysis by the finite element method', *Proc. ASCE, J. of the Hyd. Div.*, **98**, no. HY 11 (1972).

23. L. Loziuk, J. Anderson and T. Belytschko, 'Finite element approach to hydro-thermal analysis of small lakes', *Mtg. Preprint 1799, ASCE National Environmental Meeting, Houston, Texas, Oct. 1972.*

24. R. Zeller, J. Hoopes and G. Rohlich, 'Heated surface jets in steady crosscurrent', *J. of the Hyd. Div., ASCE*, **97**, no. HY 9 (1971).

25. R. Barry and D. P. Hoffman, 'Computer model for thermal plume', *Proc. ASCE, J. of the Power Div.*, **98**, no. PO 1, 117–132 (1972).

26. L. Motz and B. Benedict, 'Surface jet model for heated discharges', *Proc. ASCE, J. of the Hyd. Div.*, **98**, no. HY 1, 181–199 (1972).

27. E. Hirst, 'Zone of flow establishment for round buoyant jets', *Water Resources Res.*, **8**, 5, 1234–1246 (1972).

28. H. Stefan, C. T. Chu and H. Wing, 'Impact of cooling water on lake temperatures', *Proc. ASCE, J. of the Power Div.*, **98**, no. PO 2, 253–272 (1972).
29. H. Stefan, 'Modeling spread of heated water over lake', *Proc. ASCE, J. of the Power Div.*, **96**, no. PO 3, 469–482 (1970).
30. H. Stefan and P. Vaidayaraman, 'Jet type model for the three-dimensional thermal plume in a crosscurrent and under wind', *Water Resources Res.*, **8**, 4, 998–1014 (1972).

Chapter 7

A Finite Element Solution for Two-Dimensional Stratified Flow Problems

I. P. King, W. R. Norton and K. R. Iceman

7.1 Introduction

The modelling of thermal processes and ecological systems in reservoirs and lakes is dependent upon an adequate representation of the flow regimes. These regimes are dominated by density induced currents particularly in the region of inflows to the lake where underflow often occurs, and in the zone affected by withdrawals through outlet structures. This latter problem is particularly important because slight density variations induce considerable change in the zone of withdrawal.

The distribution and variations of wedges of saline intrusion are another area where density flows are important. This problem is significant when sites are being considered for inlet and outlet structures for power plant and industrial users. Similarly the distribution of salinity in estuaries is a continuing area of interest.

All these problems are of course truly three-dimensional but reasonable two-dimensional approximations both in vertical elevation or horizontal plan view make possible solution of a wide class of problems that are presently considered intractable.

The most satisfactory equations for flow problems are the Navier–Stokes equations which are derived on the basis of laminar flow; see References 1 and 2 for complete derivations. The extension to the turbulent situation by the so-called Reynolds approximation has a somewhat weak rational basis but does provide a method of handling turbulent effects. They have been assumed to be adequate for the ensuing model development.

The solution to any type of problem involving stratification has been very little developed. The principal efforts in this area have been made by Yih,[3] Debler[4] and Koh.[5] The problem of stratified flow into a line sink has been extensively investigated with experimental and theoretical solutions. In order to make this problem tractable, limited forms of the general equations have usually been used. A second approach to the solution of problems involving stratification has been to develop a flow field using empirical equations and solve only for temperature or other quality constituents. Such an approach has been used by Water Resources Engineers[6] for prediction of temperature in a deep reservoir.

The extension of the finite element method by restating it in terms of the Galerkin method of weighted residuals now allows a direct attack on these complex equations. This study develops a finite element formulation for both two dimensions horizontal and vertical incompressible flow systems using the Galerkin method together with the Newton–Raphson procedure for the non-linear terms.

There is relatively little data available to confirm the adequacy of computed results. This paper presents experimental data for flow over a submerged weir and compares them with computed velocities. As a demonstration of the two-dimensional horizontal model, flood flow of a river through a constriction due to bridge abutments is presented.

7.2 The fluid flow equations

7.2.1 The basic equations

The equations used below are two-dimensional versions of the general Navier–Stokes equation, the continuity equation, and the convection diffusion equations.

In accordance with conventional fluid mechanics approaches, the extension of these essentially laminar equations has been achieved by analysing the same equation in terms of mean velocities and pressures and replacing the molecular viscosity and diffusion coefficients with their turbulent counterparts. These coefficients are, of course, dependent upon the flow regime. Whilst they are assumed to be known constants for the purposes of this paper, they function at present as calibration coefficients until more experience can be achieved.

The equations as written assume that the lines of the principal diffusion coefficients coincide with the cartesian axes and that the turbulent exchange coefficients are associated with the cartesian momentum forces. In both cases, the model is easily extended to include the cross terms but the lack of data describing these coefficients does not for the moment justify a more elaborate description.

Two finite element models for fluid flow have been developed. In the first the velocities are assumed to be uniform in one horizontal direction although the thickness of the cross-section may vary, the so-called 'vertical flow' model. In the second model, velocities are assumed to be constant with depth although the depth itself may vary—the 'horizontal flow' model. Both of these developments are thus two-dimensional models.

7.2.2 Equations for vertical flow model

The equations below represent the special case developed for the assumption that velocities are uniform in the horizontal direction perpendicular to the plane of the model.

7.2.3 Momentum equations

$$b\rho\left(\frac{\partial u}{\partial t} + u\frac{\partial u}{\partial x} + v\frac{\partial u}{\partial y}\right) + \frac{\partial}{\partial x}(pb) - \varepsilon_x\left[\frac{\partial}{\partial x}\left(b\frac{\partial u}{\partial x}\right) + \frac{\partial}{\partial y}\left(b\frac{\partial u}{\partial y}\right)\right] = 0$$

(7.1a)

$$b\rho\left(\frac{\partial v}{\partial t} + u\frac{\partial v}{\partial x} + v\frac{\partial v}{\partial y}\right) + \frac{\partial}{\partial y}(pb) + b\rho g - \varepsilon_y\left[\frac{\partial}{\partial x}\left(b\frac{\partial v}{\partial x}\right) + \frac{\partial}{\partial y}\left(b\frac{\partial v}{\partial y}\right)\right] = 0$$

(7.1b)

7.2.4 Continuity equation

$$\frac{\partial}{\partial x}(bu) + \frac{\partial}{\partial y}(bv) = 0 \qquad (7.2)$$

7.2.5 Convection-diffusion equation for density

$$\frac{\partial \rho}{\partial t} + b\left(u\frac{\partial \rho}{\partial x} + v\frac{\partial \rho}{\partial y}\right) - D_x\frac{\partial}{\partial x}\left(b\frac{\partial \rho}{\partial x}\right) - D_y\frac{\partial}{\partial y}\left(b\frac{\partial \rho}{\partial y}\right) = 0 \qquad (7.3)$$

where b is the width of two-dimensional section, ε_x, ε_y are turbulent exchange coefficients, D_x, D_y are turbulent diffusion coefficients, u, v, ρ and p are as used in preceding chapters.

7.2.6 Equation for horizontal flow

When the general equations are rewritten with the assumption of uniform velocity with depth, it is appropriate to transform the pressure terms into head.

Figure 7.1 is a definition sketch which defines the relationships of water surface, bottom elevation and depth, for example

$$\frac{1}{\rho}\frac{\partial p}{\partial x} = g\frac{\partial h}{\partial x} + g\frac{\partial a_0}{\partial x} + \frac{gh}{2\rho}\frac{\partial \rho}{\partial x} \qquad (7.4)$$

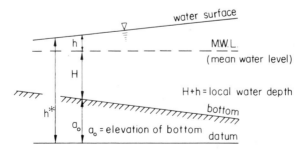

Figure 7.1 Definition sketch for horizontal flow model

where g is the acceleration due to gravity. The transformed set of equations may be written

7.2.7 *Momentum equations*

$$\frac{\partial u}{\partial t} + u\frac{\partial u}{\partial x} + w\frac{\partial u}{\partial z} + g\frac{\partial h}{\partial x} + g\frac{\partial a_0}{\partial x} + \frac{gh}{2\rho}\frac{\partial \rho}{\partial x} - \frac{\varepsilon_x}{\rho}\left(\frac{\partial^2 u}{\partial z^2} + \frac{\partial^2 u}{\partial z^2}\right) - \Lambda_w$$

$$+ \frac{gu}{C_c^2(h + H)}(u^2 + w^2)^{\frac{1}{2}} - \frac{W_x}{h} = 0 \tag{7.5a}$$

$$\frac{\partial w}{\partial t} + u\frac{\partial w}{\partial x} + w\frac{\partial w}{\partial z} + g\frac{\partial h}{\partial z} + g\frac{\partial a_0}{\partial z} + \frac{gh}{2\rho}\frac{\partial \rho}{\partial z} - \frac{\varepsilon_z}{\rho}\left(\frac{\partial^2 w}{\partial x^2} + \frac{\partial^2 w}{\partial z^2}\right) + \Lambda_u$$

$$+ \frac{gw}{C_c^2(H + h)}(u^2 + w^2)^{\frac{1}{2}} - \frac{W_z}{h} = 0 \tag{7.5b}$$

7.2.8 *Continuity equation*

$$\frac{\partial h}{\partial t} + \frac{\partial}{\partial x}[u(h + H)] + \frac{\partial}{\partial y}[w(h + H)] = 0 \tag{7.6}$$

7.2.9 *Convection-Diffusion equation for density*

$$\frac{\partial \rho}{\partial t} + u\frac{\partial \rho}{\partial x} + w\frac{\partial \rho}{\partial z} - D_x\frac{\partial^2 \rho}{\partial x^2} - D_z\frac{\partial^2 \rho}{\partial z^2} = 0 \tag{7.7}$$

where Λ is the Coriolis parameter, ε_z, D_z are turbulent exchange and diffusion coefficients, W_x, W_z are x- and z-direction surface wind shears and C_c is the Chezy coefficient.

The variable $(H + h)$ now appears in the continuity equation, making it non-linear; thus, it is convenient to transform velocities into flow quantities where $U = uH$ $V = w(H + h)$. Then, without giving complete details, substitutions may be made which lead to the final set of equations listed below.

7.2.10 *Momentum equations*

$$h^{-1}\frac{\partial U}{\partial t} - h^{-2}U\frac{\partial h}{\partial t} + Uh^{-2}\frac{\partial U}{\partial x} - U^2 h^{-3}\frac{\partial h}{\partial x} + Vh^{-2}\frac{\partial U}{\partial z}$$

$$- VUh^{-3}\frac{\partial h}{\partial z} + g\frac{\partial h}{\partial x} + g\frac{\partial a_0}{\partial x} + \frac{gh}{2\rho}\frac{\partial \rho}{\partial x}$$

$$- \frac{\varepsilon_x}{\rho}\left[h^{-1}\frac{\partial^2 U}{\partial x^2} - Uh^{-2}\frac{\partial^2 h}{\partial x^2} - h^{-2}\frac{\partial h}{\partial x}\frac{\partial U}{\partial x} + Uh^{-3}\left(\frac{\partial h}{\partial x}\right)^2\right]$$

$$- \frac{\varepsilon_x}{\rho}\left[h^{-1}\frac{\partial^2 U}{\partial z^2} - rh^{-2}\frac{\partial^2 h}{\partial z^2} - h^{-2}\frac{\partial h}{\partial z}\frac{\partial U}{\partial z} + Uh^{-3}\left(\frac{\partial h}{\partial z}\right)^2\right]$$

$$- 2V(h + H)^{-1}W + \frac{(H + h)^{-3}}{C_c^2}U(U^2 + V^2)^{\frac{1}{2}} - W_x = 0 \tag{7.8a}$$

$$h^{-1}\frac{\partial V}{\partial t} - h^{-2}V\frac{\partial h}{\partial t} + Uh^{-2}\frac{\partial V}{\partial x} - UVh^{-3}\frac{\partial h}{\partial x} + Vh^{-2}\frac{\partial V}{\partial z}$$

$$- V^2h^{-3}\frac{\partial h}{\partial z} + g\frac{\partial h}{\partial z} + g\frac{\partial a_0}{\partial z} + \frac{gh}{2\rho}\frac{\partial\rho}{\partial z}$$

$$- \frac{\varepsilon_z}{\rho}\left[h^{-1}\frac{\partial^2 V}{\partial x^2} - Vh^{-2}\frac{\partial^2 h}{\partial x} - h^{-2}\frac{\partial h}{\partial x}\frac{\partial V}{\partial x} + Vh^{-3}\left(\frac{\partial h}{\partial x}\right)^2\right]$$

$$- \frac{\varepsilon_z}{\rho}\left[h^{-1}\frac{\partial^2 V}{\partial z^2} - Vh^{-2}\frac{\partial^2 h}{\partial z^2} - h^{-2}\frac{\partial h}{\partial z}\frac{\partial V}{\partial z} + Vh^{-3}\left(\frac{\partial h}{\partial z}\right)^2\right]$$

$$+ WU(h + H)^{-1} + \frac{(H + h)^{-3}}{C_c^2}V(U^2 + V^2)^{\frac{1}{2}} - W_y = 0 \qquad (7.8b)$$

7.2.11 Continuity equation

$$\frac{\partial h}{\partial t} + \frac{\partial U}{\partial x} + \frac{\partial V}{\partial y} = 0 \qquad (7.9)$$

7.2.12 Convection-Diffusion equation for density

$$h\cdot\frac{\partial\rho}{\partial t} + U\frac{\partial\rho}{\partial x} + V\frac{\partial\rho}{\partial z} - D_x\frac{\partial^2\rho}{\partial x^2} - D_y\frac{\partial^2\rho}{\partial y^2} = 0 \qquad (7.10)$$

7.2.13 The finite element method for the flow equations

The flow equations described above differ from most conventional finite element problems in that they are highly non-linear and that there is no exact functional approach to their solution even by piecewise linearization. The development presented here attacks the equation directly and does not attempt to use stream functions or other transformations to remove the non-linearity. The Galerkin procedure in the method of weighted residuals is used to form the non-linear algebraic equations which are solved by means of the Newton–Raphson approach.

For the vertical flow equations a quadratic approximation has been chosen to represent the variations in velocities u and v and a linear variation for the pressure p and density ρ. A triangular element, with degrees of freedom as shown in Figure 7.2, is employed. This selection was made after consideration of the implication of the different order of the continuity equation. Pressure is in fact the physical significance of the Lagrange multiplier if a Lagrange multiple for the satisfaction of continuity approach is used. Linear approximations for density were chosen to simplify the already complicated computer code and space requirements.

It should also be noted that as suggested by Yih[3] and other writers, the influence of density variation on the inertia terms will be neglected and density coupling only appears through the body force term.

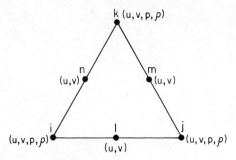

Figure 7.2   Basic element for fluid flow

As a demonstration of the element contribution, one equation for the
vertical flow model, Equation 7.1a, will be evaluated. Let the following
shape functions apply

$$u = \mathbf{Nu} \qquad v = \mathbf{Nv}$$
$$\rho = \mathbf{M\rho} \qquad p = \mathbf{Hp} \tag{7.11}$$

and assume that the width function is of the form

$$b = \mathbf{Hb} \tag{7.12}$$

After introduction of these functions into the governing equation and
their use as weighting functions in the method of weighted residuals (i.e.
using the Galerkin variation of the method of weighted residuals) one
obtains for an element

$$\mathbf{K}^v\mathbf{u} + \mathbf{D}^v\mathbf{u} + \mathbf{C}^v\dot{\mathbf{u}} + \mathbf{S}^v\mathbf{p} - \mathbf{R} = \mathbf{0} \tag{7.13}$$

where

$$\mathbf{K}^v = \iint_A \varepsilon_x[\mathbf{N}_x^T\mathbf{HbN}_x + \mathbf{N}_y^T\mathbf{HbN}_y]\,dx\,dy \tag{7.14}$$

$$\mathbf{D}^v = \iint_A \mathbf{N}^T\mathbf{Hb}\rho_A[\mathbf{NuN}_x + \mathbf{NvN}_y]\,dx\,dy \tag{7.15}$$

$$\mathbf{C}^v = \iint_A \mathbf{N}^T\mathbf{HbN}\,dx\,dy \tag{7.16}$$

$$\mathbf{S}^v = -\iint_A \mathbf{N}_x^T\mathbf{HbH}\,dx\,dy \tag{7.17}$$

$$\mathbf{R} = -\int_S (\mathbf{N}^T\mathbf{HbM}\mathbf{p} - \varepsilon_x\mathbf{N}^T\mathbf{HbN}_x\mathbf{u})\,dx$$
$$+ \int_S \varepsilon_x\mathbf{N}^T\mathbf{HbN}_x\mathbf{u}\,dy \tag{7.18}$$

in which ρ_a is an average element density (assumed constant for the system) and N_y represents the derivative with respect to y of the shape function N. It should be noted that integration by parts has been invoked in passing from the basic statement of the weighted residuals integral to the form represented by Equation 7.13.

The above operations can also be performed with similar results for the y direction of Equation 7.1b, the continuity Equation 7.2, and the advection diffusion Equation 7.3. We can denote the algebraic expressions for the complete analytical model composed of assembled finite elements as

$$\mathbf{K u}_n + \mathbf{D u}_n + \mathbf{C \dot{u}}_n + \mathbf{R} = 0 \tag{7.19}$$

where $\mathbf{u}_n$ denotes a vector of u and v velocities, pressure p and density ρ. Note that $\mathbf{D}$ is non-linear in u and $\mathbf{K}$ represents a combination of $\mathbf{K}^v$ and $\mathbf{S}^v$.

The surface integrals that develop from this process must be accounted for on boundaries of the complete system. The term associated with pressure is, in fact, the boundary pressure force distributed to the nodes. The velocity term is associated with the assumption of velocity gradients at the boundaries. If a system is analysed with specified velocity boundary conditions, or with the assumption that the boundary is away from the area of interest, this may be neglected.

The solution for time dependent problems is developed by a step by step implicit scheme in which the value of variable at the end of the time step Δt is predicted from the average time derivative over the time step, i.e.

$$u_{t+\Delta t} = u_t + \Delta t/2(\dot{u}_{t+\Delta t} + \dot{u}_t) \tag{7.20}$$

This equation and similar equations for v and ρ may be substituted into the basic equations to eliminate the time derivative and allow solution for the variables at $t = \Delta t$.

Thus the coefficient matrix of $\mathbf{C}$, Equation 7.19 may be eliminated by modifying matrices $\mathbf{K}$, $\mathbf{D}$ and $\mathbf{R}$ to incorporate time dependence. In the examples presented below only steady state solutions have been developed and thus the development for non-linear solutions will use only the steady state form.

7.3 Non-linear solution procedure

Several methods were evaluated in order to solve the non-linear equations developed by the approach outlined above. The most obvious procedures involve direct linearization of the problem by using a previous iteration value where non-linear terms occur. Two procedures are possible in this case; one using a constant coefficient matrix and the other varying the coefficient matrix for each iteration. The Newton–Raphson procedure is a third alternative. All three methods are iterative and require an estimate of the value of the unknown before solution proceeds. The direct linearization procedures proved to be unstable in a number of cases and thus the Newton–Raphson procedures was adopted for this work.

The Newton–Raphson procedure uses Equation 7.19 with a derivative approach based upon the error of a previous solution. An error function f is defined as

$$\mathbf{f}^m = \mathbf{K}\mathbf{u}_n^m + \mathbf{D}\mathbf{u}^m \mathbf{u}_n^m - \mathbf{R} \tag{7.21}$$

and solution proceeds by solving for the correction $\Delta\mathbf{u}_n$, to the previous estimate $\mathbf{u}_n$. Then

$$\mathbf{J}^m \Delta\mathbf{u}_n^{m+1} = \mathbf{f}^m \tag{7.22}$$

where

$$(J_{ij}^m) = \left(\frac{\partial f_i^m}{\partial u_j}\right)$$

and

$$\mathbf{u}_n^{m+1} = \mathbf{u}_n^m + \Delta\mathbf{u}_n^{m+1}$$

$\mathbf{J}^m$ is the so-called Jacobian of the system at step m, and terms of this matrix indicate the instantaneous slope of the functions f with respect to changes in each of the variables. Actual solution thus consists of evaluating all the error functions $\mathbf{f}$ and computing the corrections to $\mathbf{u}$ based upon the instantaneous slopes, to eliminate all the errors.

The Newton–Raphson procedure has proved to be convergent for all problems that show acceptable answers in the first step. The first step in the fluid flow equations is essentially a linear problem where a zero initial condition is assumed.

7.3.1 Application of the Newton–Raphson procedure to the fluid flow finite element

As previously stated, a typical non-linear term is $\rho_a b u\, \partial v/\partial x$ with a finite element contribution of the form

$$\mathbf{f} = \iint \mathbf{N}^{\mathrm{T}} \rho_a \mathbf{M}\mathbf{b}_n \mathbf{N}\mathbf{u}_n \mathbf{N}_{,x}\mathbf{v}_n \, \mathrm{d}A \tag{7.23}$$

The derivatives of this integral must be taken with respect to each of the variables u_n and v_n, i.e. typical terms:

$$\frac{\partial \mathbf{f}}{\partial \mathbf{u}_j} = \int \mathbf{N}^{\mathrm{T}} \rho_a \mathbf{M}\mathbf{b}_n \mathbf{N}_{,x}\mathbf{v}_n \mathbf{N}\mathrm{d}A \tag{7.24}$$

$$\frac{\partial \mathbf{f}}{\partial \mathbf{v}_j} = \int \mathbf{N}^T \rho_a \mathbf{Mb}_n \mathbf{Nu}_n \mathbf{N}_{,x} \, dA \qquad (7.25)$$

The interior matrix products $\mathbf{Mb}_n\mathbf{N}_{,x}\mathbf{v}_n$ and $\mathbf{Nu}_n$ are the local values b, $\partial v/\partial x$ and u respectively. When numerical integration is used to evaluate the integrals of Equations 7.24 and 7.25 they may be rewritten as a summation

$$\frac{\partial \mathbf{f}}{\partial \mathbf{u}_j} = \rho_a A \sum w_e \mathbf{N}^T b_e \left(\frac{\partial v}{\partial x}\right)_e \mathbf{N} \qquad (7.26)$$

$$\frac{\partial \mathbf{f}}{\partial \mathbf{v}_j} = \rho_a A \sum w_e \mathbf{N}^T b_e u_e \mathbf{N}_x \qquad (7.27)$$

where A = area of triangle, w_e = numerical integration weighting factor at location e, and e = indicates evaluation at the specified integration point.

Thus, if the four integrals derived from the Galerkin approach for momentum continuity and convection diffusion, are listed as f_1, f_2, f_3 and f_4 respectively, the triangular element coefficient matrix takes the form

$$
\begin{array}{c}
\quad 6 \quad\;\; 6 \quad\;\; 3 \quad\;\; 3 \\
\begin{array}{c} 6 \\[18pt] 6 \\[18pt] 3 \\[18pt] 3 \end{array}
\begin{bmatrix}
\dfrac{\partial f_1}{\partial u_j} & \dfrac{\partial f_1}{\partial v_j} & \dfrac{\partial f_1}{\partial p_j} & 0 \\[10pt]
\dfrac{\partial f_2}{\partial u_j} & \dfrac{\partial f_2}{\partial v_j} & \dfrac{\partial f_2}{\partial p_j} & \dfrac{\partial f_2}{\partial \rho_j} \\[10pt]
\dfrac{\partial f_3}{\partial u_j} & \dfrac{\partial f_3}{\partial v_j} & 0 & 0 \\[10pt]
\dfrac{\partial f_4}{\partial u_j} & \dfrac{\partial f_4}{\partial v_j} & 0 & \dfrac{\partial f_4}{\partial \rho_j}
\end{bmatrix}
\begin{bmatrix} \Delta u_j \\[10pt] \Delta v_j \\[10pt] \Delta p_j \\[10pt] \Delta \rho_j \end{bmatrix}
-
\begin{bmatrix} F_1 \\[10pt] F_2 \\[10pt] F_3 \\[10pt] F_4 \end{bmatrix}
= 0 \qquad (7.28)
$$

Note that the zero occurs where an integral is independent of that variable. A complete list of all terms in this matrix is given in Reference 8.

Formulation for the horizontal flow equation follows exactly the same procedure except that pressure is replaced as a variable by head and because the equations are more complex the terms become very large and cumbersome.

7.3.2 The computer model

As previously stated two computer models have been developed. These models are essentially the same in form using the same equation structure and process and having the same general input and output requirements. The main differences, of course, reside in the element coefficient generation routine.

An in-core equation solver has been used throughout to reduce run times on the several iterations required to produce even a steady state solution. The equations are, of course, unsymmetrical; this approximately doubles computer storage requirements and quadruples solution time over similar problems with symmetrical equations. The solution routine uses a dynamically allocated storage algorithm with a pseudo-rectangular form in which the diagonal is located to minimize the storage actually used. One other feature of the model allows the user to, at his option, control the order of the solution process.

The 'vertical flow' model must also allow variation in the free surface. The solution to this problem depends on whether the system to be analysed is transient or in steady state. For transient situations, flow is allowed to move in or out across the surface and then continuity conditions are used to adjust the surface element and the adjacent elements. The steady state system poses a different problem in that the final surface elevation is desired. This is achieved by solving problems with restraint against velocity normal to the surface. Such restraint induces a pressure at the surface. This pressure is then used to develop the final surface profile by a process of coordinate adjustment with each step of the iteration.

7.4 Test problem

There are many facets of the modelling of finite element flows which require evaluation before a potential user can have confidence in application of the model. Four of the questions will be discussed with reference to a simple test problem. The four topics are:

(1) Influence of network detail.
(2) Specification of boundary conditions at corners.
(3) Non-linear solution convergence.
(4) Influence of coefficients.

A single example problem will be used as a reference in this discussion. The problem of flow around a 90° bend has been selected, the dimensions of the test problem is shown in Figure 7.3 together with the location of cross-sections for the later discussion.

7.4.1 *Influence of network detail*

Network details, of course, one of the fundamental and difficult problems faced by users of the finite element method. Theoretical discussions of convergence have been engaged in by many researchers and it is not the intention of this paper to follow this line. Instead, the variation of the solution for different network detail will be described. The three test networks shown

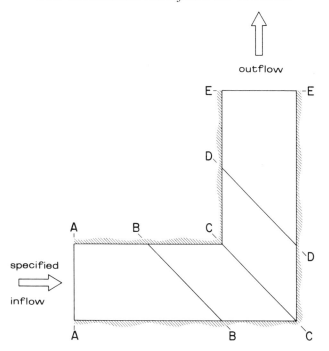

Figure 7.3 Horizontal flow around a 90° bend test problem to show the influence of finite element network detail

in Figure 7.4 were subject to identical upstream flow and downstream head conditions, the turbulent exchange coefficients were set equal to 100 lb. s/ft². Figure 7.5 shows the velocity profiles for differing cross-sections for all three networks.

The velocities shown at Section AA are those specified by input and reflect a uniform distribution which is essentially identical for each network. The difference reflects different water depths. At Section BB the velocities are nearly identical but as the flow reaches the bend itself, Section CC, there is considerable difference between the results of Network I and those predicted by Network II and III. Significantly, however, the results for Networks II and III are very similar. The velocity profiles at Section DD are strongly influenced by the amount of detail at the bend, thus the two more refined networks show similar behaviour which is in turn very different from Network I. At the system exit, Section EE, the influence of the bend is decreasing and the profiles become similar again.

Figure 7.6 demonstrates a further basis for comparison of results from the three networks—continuity checking. Total flows crossing the specified sections are shown for all three networks. The nature of the method ensures

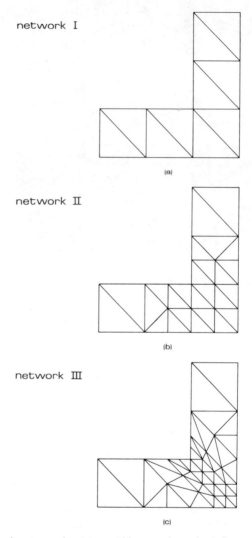

Figure 7.4 Network constructions test problem to show the influence of finite element network detail

total flow continuity across external boundaries but not on internal sections. Figure 7.6 reveals the steady improvement in the satisfaction of continuity as the network becomes finer. Network I shows a maximum error of 28 per cent and fairly poor checks at sections downstream of the bend. Network II shows improvement with a 15 per cent error at the bend and less than 8 per cent elsewhere and fault. Network III has a maximum error of 7 per cent and less than 2 per cent at other sections.

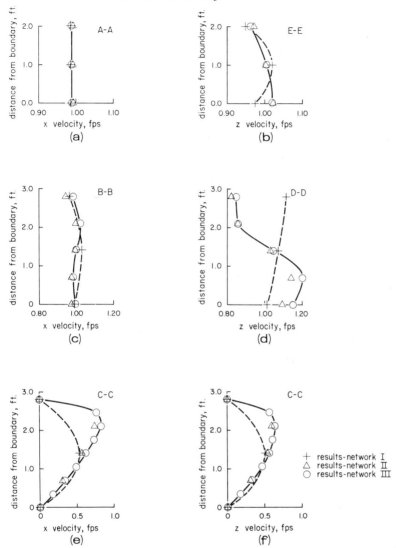

Figure 7.5 Comparison of velocity profiles for three different network constructions

These results lead to the broad conclusion, common to many finite element procedures, that in areas of rapid gradients more detail is required but away from the detail problem quite coarse networks may be used. The similarity of solutions for Networks II and III at the downstream section suggest that areas of where there is little interest may be approximated by quite coarse networks without deteriorating results elsewhere.

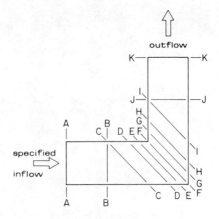

location sketch for lines of
continuity checking

(a)

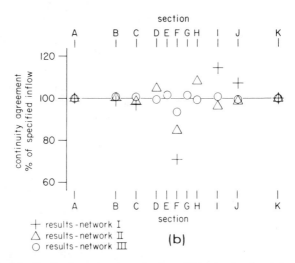

+ results - network I
△ results - network II
○ results - network III

(b)

Figure 7.6 Continuity comparisons for three different network constructions

7.4.2 *Specification of boundary conditions at corners*

One of the most troublesome problems encountered in the application of
the finite element method models to fluid flow has been the proper specification of boundary conditions at breaks in the straight line boundary which
confines the fluid system.

The problem arises directly as a result of the fact that the only proper
velocity specification at a corner is zero in both the X and Y or X and Z

direction. If a velocity vector, other than zero, is allowed to exist at a corner it cannot be parallel to the direction of both boundaries entering that corner. Such a vector would then imply flow across one or both the boundaries. In the mathematical formulation the primary reason such a zero specification causes trouble is that it forces the simulation to meet this specification at the corner to the detriment of some other area in the problem. An example of this problem is indicated in Figure 7.5 at Section CC. Here it can be seen that the zero velocity specifications at both corners leave the finite element method very little room to manoeuvre and thus force certain inaccuracies in the overall results.

This problem can be minimized by structuring networks which have relatively more detail near corners and thus have more degrees of freedom and more flexibility in meeting the severe demands imposed by these artificial boundary conditions. This approach, plus the specification of differential turbulent exchange coefficients, has been employed in the model-prototype comparisons and the example problem applications. Such constructions have proved reasonably satisfactory for this work, but the writers believe that a more formal resolution of the problem of steep gradients at corners is essential.

7.4.3 Non-linear solution convergence

The models developed in this paper include several non-linear terms in the motion equations. The inclusion of these terms precludes the direct and exact solution of overall system equations by any presently known numerical method and iteration must be used.

It should be noted that slow convergence or a lack of convergence when iterating may be caused by several things. Slow convergence is usually caused by a bad initial estimate of the solution or extreme sensitivity to small changes in certain system parameters. Also since the iterative scheme works on the non-linear terms, the magnitude of those terms, relative to other terms, will often influence the speed of solution convergence. It has been the writers' experience that simulations of homogeneous flow problems converge to satisfactory solutions much faster than non-homogeneous flow problems. In most homogeneous cases the maximum change in system state variables after about four iteration cycles (for steady state or the initial time step) has usually been in the order of less than 1 per cent. The next most rapid solutions have been obtained for non-homogeneous horizontal flow problems whose convergence differs very little, if any, from the homogeneous cases. The only really difficult convergence problems have been encountered in the non-homogeneous vertical flow simulations with very low (laminar range) velocities. In these problems it is not uncommon to have very slow convergence due to sensitivity of the solution to small changes in the pressure and density gradients. There is also a point where the precision of the computer doing the numerical work may become a factor. In none of the results shown here is precision a known problem.

7.4.4 Influence of turbulent exchange coefficients

It is important to recognize that the terms containing the turbulent exchange coefficients are simply one of several and that their actual importance is related directly to their relative magnitude in the prototype system. What this implies is that for problems in which behaviour is known to be largely dominated by the inertial terms one need be less concerned by the exact specification of exchange coefficients than in problems where the viscous or turbulent exchange terms are known to be relatively more important. On the other hand, if the inertial terms are known to be relatively small, we are again in a fairly good position, since the overall effects of turbulence will also be less, and exchange coefficients from the literature can be used. The most difficult problem arises in the case where the inertial and viscous forces are of roughly equal magnitude, and proper coefficient specification becomes essential to accurate modelling. The frequency with which this situation will be encountered is not known, but present indications are that most prototype problems will be dominated by the inertial terms and thus adequate modelling will not await additional research on these coefficients. For these cases nominal values of the various coefficients can be specified with reasonable assurance that accurate answers will result.

A review of available literature suggests that the best estimates for turbulent exchange coefficients come from the work in oceanography.[7] Even here there are great gaps, however, as various problems in measurement techniques have hindered the determination of a reliable set of numbers based on bulk flow characteristics. All that can be said with certainty is that the magnitude of the coefficients is scaled by at least the following factors: depth, velocity and velocity gradient, scale of motion, degree of stratification and relative time scale.

Of the values reported in the literature the most consistent numbers seem to be in the order of $0–10^6$ lb. s/ft^2 for turbulent exchange. The diffusion coefficients should be of roughly the same order only smaller. An assessment of the Richardson number may be useful in determining the ratio between the exchange coefficients.

Experience has shown that the appropriate coefficients to use in the finite element method models for reservoir type problems are very much lower than those given usually in the literature. For the problems that have been simulated thus far, the best values for use in the horizontal plane have been found to be in the range of 10^{-3} to 10^2 and in the vertical plane from 10^{-3} to 10^1. Table 7.1 summarizes the exchange coefficients used in various examples that have been tested.

Table 7.1 Values of exchange coefficients used in example problems

Type of simulation	Values of turbulent exchange coefficients (eddy viscosity) lb. s/ft^2			Values of turbulent diffusion coefficients (eddy diffusivity)		
	ε_x	ε_y	ε_z	D_x	D_y	D_z
Homogeneous flow over a broad-crested weir—turbulent range	0·035	0·070		—	—	—
Non-homogeneous flow over a broad-crested weir—laminar range	0·001 to 0·100	0·001 to 0·100		0·100	0·0001	—
Homogeneous horizontal flow around an island—turbulent range	10 to 100		10 to 100	—	—	—
Homogeneous horizontal flow at a confluence—turbulent range	25 to 100		25 to 100	—	—	—
Homogeneous vertical flow at a river confluence—turbulent range	10	10	—	—	—	—
Non-homogeneous vertical flow over a river confluence—turbulent range	10	10	—	1·0	1·0	—
Homogeneous horizontal flow of a river in flood	50 to 750		50 to 750	—	—	—

The causes of the differences between reported coefficients and those found most usable for the models is not fully known, but are believed to stem primarily from three sources. First, since most values reported in the literature have come from the study of very large lakes and the open ocean it is quite possible that influence of 'scale of motion' is simply enough to produce the large numbers reported and thus the use of smaller problems is valid. A second likely explanation for the discrepancy is simply the difficulty in measuring the numbers. In order to determine values for these coefficients one must assemble a large group of people, disperse them spatially with precise coordination, and supply them with extremely accurate and consistent measuring devices. It is not entirely clear that this can be done to the level of precision required for the problem. A final possible explanation, and the one that seems most likely to us, is that the models used to determine the reported values were not as complete as those reported in this work and that influence of important terms has been inadvertently 'lumped' into the turbulent exchange term. Should this be the case one could expect to use different coefficients in the finite element method models than those reported.

7.5 Demonstration problems

As examples of the use of the models in practical situations the analysis of two problems will be described. The vertical flow model will be demonstrated by analysis of flow over a submerged weir and the horizontal flow model by river flow at flood levels through a constricted gap caused by a highway bridge.

7.5.1   Flow over a submerged weir

All the data described for this model–prototype comparison were collected in a laboratory flume at the U.S. Corps of Engineers Waterways Experimental Station. The velocity measurements were made in the vicinity of a broad-crested weir near the downstream 14 ft end of a one foot wide rectangular flume which was more than 50 ft long. The weir which had a height of 1·2 ft was located 7 ft from the end of the flume. The comparison is presented for two cases of two-dimensional steady flow in the vertical plane.

Case 1 : Homogeneous flow with a Reynolds number of approximately 70 000.
Case 2 : Stratified flows with a Reynolds number of approximately 7400 and two layers of fluid with differing densities.

Figure 7.7 shows the network used in Case 1. The upstream velocity profile was fixed and exit pressures were known to be hydrostatically

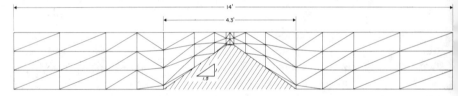

Figure 7.7   Finite element network for weir simulation

distributed. The turbulent exchange coefficients ε_x, ε_y were given constant values of 0·035 and 0·070 lb. s/ft^2 and the solution was repeated through six iterations until satisfactory convergence (velocity changes of less than 0·001 ft/s) was achieved. Figure 7.8 shows a comparison of simulated and observed velocities both upstream and downstream of the weir. The observed profiles in the upstream section show excellent agreement. Downstream, only the exit profile was observed but a qualitative sketch of an eddy was given, the simulated results show correspondence with this area. The model was calibrated by adjusting the exchange coefficients until the water surface profile showed good correspondence. Figure 7.9 shows the final water surface profile. The error is probably associated with the lack of detail at the weir crest, and a finer mesh would probably reduce this discrepancy. Notice that the upstream slope is consistent with observed values.

Case 2 : The upstream condition consisted of the velocities shown with water of specific gravity 1·007 for the bottom 0·35 ft and 1·000 for the top 130 ft. The element network was refined somewhat in the area of sharp

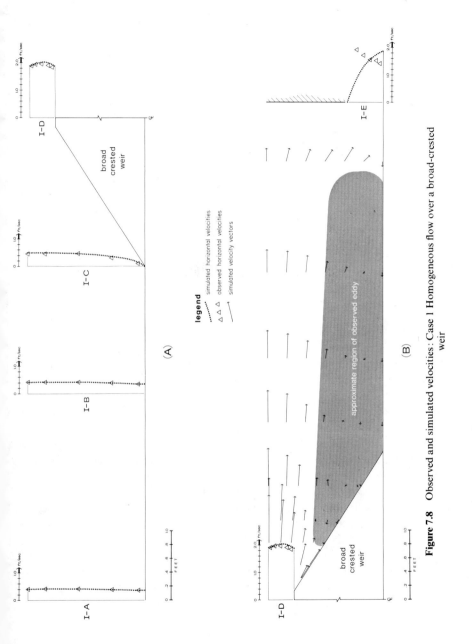

Figure 7.8 Observed and simulated velocities: Case 1 Homogeneous flow over a broad-crested weir

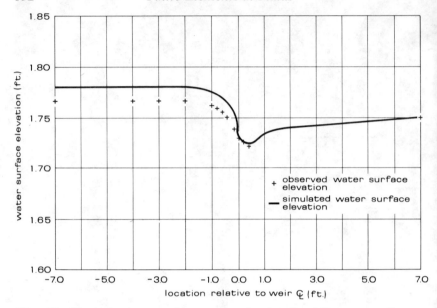

Figure 7.9   Observed and computed water surface elevation: Case 1 Homogeneous flow over
a broad-crested weir

density gradients but was essentially the same as shown in Figure 7.7 for
Case 1. The turbulent exchange coefficients varied between 0·01 and 0·0001
lb. s/ft^2 depending on the location in the fluid body. These values were set in
attempting to match the shape of velocity profile on the upstream section.
Downstream, fairly homogeneous flow was indicated and a constant value
of 0·004 lb. s/ft^2 were used. The diffusion coefficients were set equal to
1·000 ft^2/s for the whole system.

Figure 7.10 shows a comparison of observed and simulated velocity
profiles. The band indicated for the observed profiles was caused by plotting
measured values at various points in the cross-section (i.e. 0·1 ft, 0·25 ft and
0·5 ft on side wall). The inconsistencies of these measurements make it
appropriate to plot the measured values as an area. There is generally
acceptable agreement between the simulated and observed profiles. The
downstream eddy shows consistently good agreement and the withdrawal
profile is satisfactory.

There is some numerical dispersion in the area where stratified flow
occurs; part of this is probably caused by the difficulty in specifying unique
boundary conditions close to the foot of the weir.

7.5.2   River flow at a bridge abutment

As a demonstration of the horizontal flow model, a simulation has been made
of flood flow of a river through the constriction caused by a road bridge.

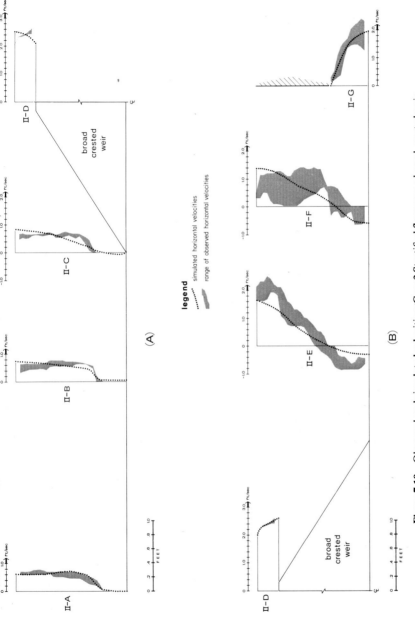

Figure 7.10   Observed and simulated velocities : Case 2 Stratified flow over a broad-crested weir

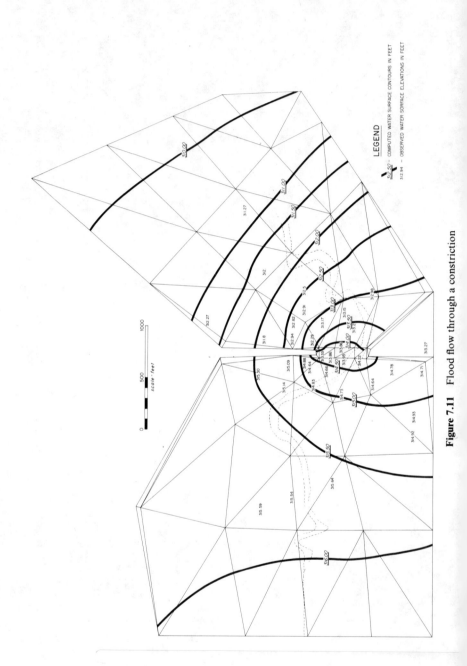

Figure 7.11 Flood flow through a constriction

The network used is shown as background in Figure 7.11; the total flow modelled is 20 000 ft³/s and much of the area covered was scrubland. The river bed which is shown as a dotted line represents only a small segment of the area. Values of exchange coefficients were in the range of 50–500 lb. s/ft².

A comparison of measurements made in the field and the simulated results is also shown in Figure 7.11. The contours, which are drawn from computed results, and measured values are shown at points. The agreement is generally very good except close to the bridge opening. In this region the lack of detail close to the opening causes the depression due to the constriction effect to be underestimated.

7.6 Concluding remarks

The computer model has been developed to simulate density stratified flow for both vertical and horizontal systems. The model has been tested on a simplified system to improve the understanding of the model and build experience with coefficients. Data is so sparse for density stratified flow problems that only a limited demonstration is possible for comparison with prototype data.

Current research is aimed at building experience with use of the models and understanding their limitations. Further consideration is being given to incorporating isoparametric elements to simplify some of the problems created by sharply changing boundary configurations.

7.7 Acknowledgements

The writers wish to acknowledge the Office of Water Resources Research, U.S. Department of the Interior, and U.S. Army Corps of Engineers, Walla Walla, who supported the research described in this paper. The assistance of Dr. J. Grace, Waterways Experimental Station, Mississippi, who provided the data for the flow over the weir is also acknowledged.

References

1. H. Lamb, *Hydrodynamics*, Dover Publications, New York, 1945.
2. B. Le Mehaute, *An Introduction to Hydrodynamics and Water Waves*, U.S. Department of Commerce ESSA Report ERL 118 Pol 3–1, July 1969.
3. C. S. Yih, *Dynamics of Nonhomogeneous Fluids*, MacMillan, New York, 1965.
4. W. R. Debler, 'Stratified flow into a line sink', *Journal of the Engineering Mechanics Division, ASCE*, **85**, No. EM3, 51–65 (1959).
5. R. C. Y. Koh, *Viscous Stratified Flow Toward a Line Sink*, California Institute of Technology Report No. KH-R-6, January 1964.
6. Water Resources Engineers, Inc., *Mathematical Models for the Prediction of Thermal Energy Changes in Impoundments*, Water Quality Office EPA, Report No. 16130.

7. H. U. Sverdrup, M. W. Johnson and R. H. Fleming, *The Oceans, Their Physics, Chemistry, and General Biology*, Prentice-Hall, 1942.
8. I. P. King, W. R. Norton and G. T. Orlob, *Finite Element Model for Two-Dimensional Density Stratified Flow*, Water Resources Engineers, Inc. Report to OWRR, 1973.

Chapter 8

Finite Element Methods for Flow in Porous Media

C. S. Desai

8.1 Introduction

The phenomenon of fluid flow or seepage through porous media is encountered in various disciplines of engineering science (see Table 8.1). The primary purpose of this chapter is to present a state-of-the-art report on the development and application of the finite element (FE) method.

Table 8.1 Fields of applications

Field	Topics
1. Water resources development and hydrology	Ground water flow; aquifer analysis and performance; saturated and unsaturated flow; recharge, evaporation and transpiration
2. Geotechnical engineering (soil and rock) mechanics	Seepage through foundations of dams, sheet piles and cofferdams, and through earth and rock-fill banks and tidal beaches; flow through jointed (rock) media; flow towards underground openings.
	Coupled flow—consolidation and expansion or swelling, liquefaction, electro-osmosis.
3. Irrigation and drainage	Flow towards and from canals, ditches, drains, wells, and through dams.
4. Soil science	Moisture movements through saturated and partially saturated soils; flow towards plant roots.
5. Hydraulics	Flow—Darcy and non-Darcy—through earth and rock-fill dams and foundations, and banks of waterways.
6. Oil reservoir engineering	Production of oil and gas.
7. Environmental engineering	Dispersion of contaminants through soils and rocks; underground disposal of liquid wastes.

Before the era of the FE method, the finite difference (FD) method was the main and most widely used numerical procedure. Comprehensive reviews and lists of annotated references for the FD method in the subject topic are given by Desai[1] and Remson and coworkers.[2] In the following, the FD method is stated briefly in context only, and some of the characteristics of the two methods are compared on the basis of typical solved problems.

The main subject that we shall consider is seepage through porous rigid media; such problems as consolidation, expansion or swelling, liquefaction and electro-osmosis and diffusion will be reviewed very briefly.

8.2   Statement of applications of the FE method

Pioneering work towards the development and use of the FE method in the subject area has been done by Zienkiewicz and coworkers.[3–6] Details of formulations and comprehensive reviews have been presented by Zienkiewicz,[6] Desai and Abel[7] and Desai.[1,8,9] The developments and applications in the past have considered various categories of flow—steady confined,[3,6,7,9–14] steady unconfined or free surface,[6,7,9,12,13,15–22] unsteady or transient confined[4,23–26] and transient unconfined.[6,7,27–41] A description of these categories of seepage will be presented subsequently.

A number of investigations have considered the problem of flow through porous jointed or discontinuous media;[42–46] the effect of fluid pressure on the deformation of jointed media has been accounted for by using iterative procedures.[45] For a comprehensive review of material properties of jointed media and for a list of references, the reader can consult a recent proceedings volume.[46]

The FE method has been formulated for the solution of diffusion–convection equation.[47–51] The coupled problems, typified by consolidation,[52–58] liquefaction[59] and electro-osmosis,[60,61] have been formulated and preliminary applications have been made. Applications to various problems in the foregoing topics are also presented in the proceedings of the conference from which this volume stems.[62]

The FE method has been used for the problem of flow through porous media almost from its inception. The literature is scattered in publications of various disciplines. Hence, although an effort has been made to review a number of publications, owing to limitations on space and accessibility, some significant contributions are bound to have been missed.

8.3   Theoretical aspects

The formulations for the flow problem are based on a number of assumptions. Chief among these assumptions are that the medium is rigid and continuous, the fluid is homogeneous and incompressible, flow is continuous and irrotational, and capillary and inertia effects are negligible. Many formulations have been based on Darcy's law;[63,64] some of the works have developed procedures based on non-Darcy laws.[65,67] The Dupuit assumption has often been employed to obtain simplified formulations.

The flow problem is formulated by constructing theoretical models; the two types of models used are deterministic and probabilistic.[21,68–72]

The FE applications have been essentially based on the deterministic models in which the problem is commonly dealt with on the basis of two concepts: *particulate* and *continuum*.[21,72] In the particulate approach, idealized physical models are used to define flow behaviour and mathematical expressions are established in the continuum approach. Most FE formulations are based on the latter concept in which the porous medium—an assemblage of solid particles, voids and fluids—is replaced by an equivalent continuum. The material properties are determined on the basis of (laboratory) tests on samples of the continuum. We shall cover herein details of the continuum approach.

8.3.1 Constitutive laws

The Darcy's law assumed in many developments can be expressed as

$$v = -k\phi' \tag{8.1}$$

where v = average seepage velocity, ϕ' = hydraulic gradient, and k = coefficient of permeability in the direction of flow. As an approximation, k is often assumed to be a constant.

Darcy's law can be considered to be valid for creeping flows with very low values (nearly equal to zero) of Reynolds number Re, and for flow with $Re < 1$. The latter can be assumed to occur in media with particle size less than 1 mm.[21] Its validity is dependent upon the properties of the porous medium as, for instance, its use can be justified for flow through materials like clayey and silty soils and silty fine sands. The extent of the flow regime thus covered by the Darcy's law will involve the pre-laminar regime, and a part of the laminar regime with small magnitudes of fluid velocity wherein the effects of inertia terms in the Navier–Stokes equations can be ignored in comparison with those of the viscosity terms.[21]

The flow behaviour is essentially non-linear in the laminar regime and beyond a value of $Re < 1$ it may be necessary to use non-linear or non-Darcy laws. Two non-linear laws used in the FE formulations are

$$\phi' = av + bv^2 \tag{8.2a}$$

and

$$\phi' = cv^m \tag{8.2b}$$

where a, b, c and m = material constants that can be determined from (permeameter) tests. The relation in Equation 8.2a is called the Forchheimer law[65] with a and b assumed to be invariant, and that in Equation 8.2b is known as Missbach or exponential law.[66] The Forchheimer relation can be assumed to be valid for various ranges of $Re = 0–5, 5–25, 25–100$ within the laminar regime, and it has been observed by Trollope, Stark, and Volker[21] that Equation 8.2a can also be extended into the turbulent flow regime. The turbulent flow regime needs to be considered at high Re.

In the FE and FD methods, use of the non-linear equations, (8.2a) and (8.2b), have been reported by Fenton,[17] Volker,[18] McCorquodale,[28] Trollope, Stark and Volker[21] and Chowdhury.[62]

8.3.2 Governing equations

A governing equation for saturated flow used in the FE formulations can be expressed as[6,7]

$$\frac{\partial}{\partial x}\left(k_x\frac{\partial \phi}{\partial x}\right) + \frac{\partial}{\partial y}\left(k_y\frac{\partial \phi}{\partial y}\right) + \frac{\partial}{\partial z}\left(k_z\frac{\partial \phi}{\partial z}\right) + \bar{Q} = n\frac{\partial \phi}{\partial t} \qquad (8.3)$$

where k_x, k_y, k_z = coefficients of permeability (in units of length divided by time) in the x, y and z directions, respectively, n = effective porosity divided by aquifer thickness (in units of 1/length) for unconfined flow and specific storage (also in units of 1/length) for confined flow, t = time, $\phi = p/\gamma + z$ fluid head or potential, p = pressure, γ = unit weight, z = elevation head and $\bar{Q}$ = applied fluid flux. An equation governing partially saturated or unsaturated flow may be expressed as[2,73]

$$\frac{\partial}{\partial x}\left[\rho k(\theta)\frac{\partial \phi}{\partial x}\right] + \frac{\partial}{\partial y}\left[\rho k(\theta)\frac{\partial \phi}{\partial y}\right] + \frac{\partial}{\partial x}\left[\rho k(\theta)\frac{\partial \phi}{\partial z}\right] = \frac{\partial(\rho\theta)}{\partial t} \qquad (8.4)$$

where θ = moisture content expressed as volume fraction and ρ = density of the fluid.

An equation, called the Boussinesq equation,[27,36,38,74] governing one-dimensional flow often solved by using the FE and FD methods is

$$k_x\frac{\partial}{\partial x}\left(h\frac{\partial h}{\partial x}\right) = n\frac{\partial h}{\partial t} \qquad (8.5)$$

where h = height of the free surface. It is possible to use linearized versions of Equation 8.5 and similar versions of two-dimensional equations.[27,35,38,74,75]

A form of the governing equation for non-Darcy flow, given by McCorquodale[28] is

$$\frac{\partial}{\partial x_i}\left[k(v)\frac{\partial \theta}{\partial x_i} + \frac{1}{gn}\frac{\partial v_i}{\partial t}\right] = 0 \qquad (8.6)$$

in which g = acceleration due to gravity, x_i = coordinate axes ($i = 1, 2, 3$) and $k(v)$ = hydraulic conductivity.

8.3.3 Categories of seepage

Equations 8.3 and 8.6 represent *unsteady* or *transient* flow; if the time dependent term vanishes, we obtain the equation for *steady* seepage. Some common subcategories of seepage are *confined* and *unconfined* or *free surface* (FS).

Examples of steady confined, transient confined, steady unconfined, and transient unconfined flow are shown in Figures 8.1(a), (b), (c) and (d) respectively.

8.3.4 Boundary conditions

The common boundary conditions occurring in the flow problem can be expressed as (Figure 8.1)

$$\phi = \bar{\phi}(t) \quad \text{on } \Gamma_1 \tag{8.7a}$$

$$k_x \frac{\partial \phi}{\partial x} l_y + k_y \frac{\partial \phi}{\partial y} l_y + k_z \frac{\partial \phi}{\partial z} l_z + \bar{q}(t) = 0 \quad \text{on } \Gamma_2 \tag{8.7b}$$

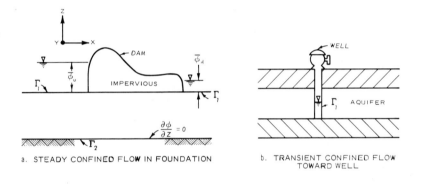

a. STEADY CONFINED FLOW IN FOUNDATION

b. TRANSIENT CONFINED FLOW TOWARD WELL

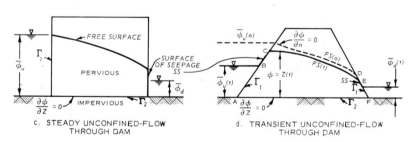

c. STEADY UNCONFINED-FLOW THROUGH DAM

d. TRANSIENT UNCONFINED-FLOW THROUGH DAM

Figure 8.1 Categories of flow and boundary conditions

where $\bar{\phi}(t)$ = prescribed potential on the boundary Γ_1, Figure 8.1, l_x, l_y, l_z = direction cosines of the outward normal to the boundary Γ_2 on which the flow is prescribed, and $\bar{q}(t)$ = prescribed intensity of flow.

The condition in Equation 8.7a is called the potential or head boundary condition, and that in Equation 8.7b is called the flow boundary condition. These two conditions, respectively, represent Dirichlet and Neumann type boundary conditions. A prescribed variation of fluid head on the upstream face of the dam, Figure 8.1(c), is an example of the potential boundary condition and vanishing of flow across the impervious base; Figures 8.1(a), (c), (d) represent the flow boundary condition.

In the case of unconfined flow, additional (mixed) boundary conditions need to be satisfied. For instance, in transient unconfined flow caused by the time dependent fluctuations in the reservoir, Figure 8.1(d), the boundary conditions are

$$\phi = \bar{\phi}_u(t) \quad \text{at upstream face AB} \tag{8.8a}$$

$$\phi = \bar{\phi}_d(t) \quad \text{at downstream face EF}$$

$$\frac{\partial \phi}{\partial z} = 0 \quad \text{along the impervious base AF} \tag{8.8b}$$

$$\frac{\partial \phi}{\partial n} = 0 \quad \text{along the free surface CD} \tag{8.8c}$$

$$\phi = z(t) \quad \text{along the free surface CD} \tag{8.8d}$$

$$\phi = z(t) \quad \text{along the surfaces of seepage BC and DE}$$

where $z(t)$ = vertical coordinate and n denotes normal to the boundary.

8.4 Formulation procedures

Variational and weighted residual methods have been employed commonly for formulating FE equations. These procedures are covered fully in Chapters 2 and 4 and elsewhere in this book.

Variational procedures were primarily used in the initial stages of the FE method; we shall state here some of these procedures. The variational functional used commonly in the earlier applications can be expressed as

$$\Pi_p(\phi) = \int_\Omega \frac{1}{2} \left[k_x \left(\frac{\partial \phi}{\partial x} \right)^2 + k_y \left(\frac{\partial \phi}{\partial y} \right)^2 + k_z \left(\frac{\partial \phi}{\partial z} \right)^2 \right.$$
$$\left. - 2 \left(\bar{Q} - n \frac{\partial \phi}{\partial t} \right) \phi \right] d\Omega - \int\int_{\Gamma_2} \bar{q}\phi \, d\Gamma \tag{8.9}$$

in which Ω = domain (volume) of flow, Γ = surface, Γ_2 = part of the surface (area) on which $\bar{q}$ is prescribed. Alternative functionals in terms of the stream function, ψ, can also be employed; Meissner[62] has presented a set of generalized variational principles for the flow problems.

Variational functionals based on the Gurtin's principle[76] have been formalized by Sandhu and Wilson[52] for the consolidation problem, and by Javandel and Witherspoon[23,24] and Neuman and Witherspoon[19,30,77] for the flow through rigid and slightly compressible media. A form of such a functional given by Neuman and Witherspoon is

$$\Pi_p(\phi, h) = \int_\Omega \left(\tfrac{1}{2} k_{ij} \frac{\partial \phi}{\partial x_j} \frac{\partial \phi}{\partial x_i} + S_s \phi \frac{\partial \phi}{\partial t} \right) d\Omega - \int\int_{\Gamma_1} (\phi - \bar{\phi}) k_{ij} \frac{\partial \phi}{\partial x_j} n_i \, d\Gamma$$

$$+ \int\int_{\Gamma_2} \bar{q} \phi \, d\Gamma - \int_{FS} (\phi - h) k_{ij} \frac{\partial \phi}{\partial x_j} n_i \, d\Gamma - \int_{FS} h \left(\bar{q}_F - S_y \frac{\partial h}{\partial t} \right)$$

$$\times n_3 \, d\Gamma + \int_{SS} (\phi - x_3) k_{ij} \frac{\partial \phi}{\partial x_j} n_i \, d\Gamma \tag{8.10}$$

where k_{ij} = permeability tensor, n_i = unit outward normal vector, FS = free surface, SS = surface of seepage, S_s = specific storage and S_y = specific yield of the medium.

Satisfactory solutions have been obtained by using the functional in Equation 8.9. It has been observed, however, that such a procedure for time dependent problems does not fulfil the requirements of the calculus of variations. This is because during variations of ϕ, $\partial \phi / \partial t$ is assumed to remain invariant. This principle is, therefore, referred to as pseudo-variational principle. Details of true, quasi- and restricted variational principles are given by Finlayson[78] and are discussed by Aral and coworkers[79] and Desai.[8]

Variational formulations with the FE method for non-Darcy flow have been reported by Fenton,[17] Volker[18] and McCorquodale.[28] The variational functional for two-dimensional flow corresponding to Equation 8.6 is

$$\Pi(\phi) = \int_\Omega \left\{ k(|\nabla \phi|) \left(\left(\frac{\partial \phi}{\partial x} \right)^2 + \left(\frac{\partial \phi}{\partial y} \right)^2 \right) \right.$$

$$\left. + \frac{a}{6bc'} (2 - c'|\nabla \phi|) \sqrt{1 + c'|\nabla \phi|} \right\} dx \, dy \tag{8.11}$$

where $c' = 4b/a^2$. The second part in the functional, Equation 8.11, was introduced to maintain continuity of flow.

Weighted residual procedures, particularly the Galerkin residual method, have been found to be more general for the flow problem which is governed by non-linear equations and the reader will find these employed extensively in the other chapters of this book.

8.5   Treatment of boundary conditions

8.5.1   Free surface

Boundary conditions in the form of prescribed heads and non-zero quantity of flow can be dealt with easily in the finite element formulation. More difficult circumstances arise when there is a free surface or under conditions of saturated–unsaturated flow.

Free surface for steady flow is determined usually by using iterative procedures.[9,15–22,29] The location of the free surface is first guessed and then modified successively on the basis of the values of fluid heads computed at each step of iteration. The boundary conditions of no flow across and of the total head equal to the elevation head at the FS, Equations 8.8c and 8.8d, are verified at each step, and the procedure is carried out until the movements of the FS become essentially negligible.

Location of the transient free surface is relatively more difficult. A number of procedures have been proposed. The procedure proposed by Desai[9,27,33] and Taylor and coworkers[29] is based on the solution of the steady state version of Equation 8.3. At each time interval, the movements of the free surface (nodes) are obtained from computed values of nodal heads and velocities. The coordinates of the nodes on the FS are then modified as

$$x_i^{t+\Delta t} = x_i^t + \Delta t \dot{x}_i(t')$$ (8.12)

where x_i denotes coordinates of nodal points on the FS, $i = 1, 2 \ldots N$, N = number of nodes on FS, t = time, Δt = time increment, the overdot denotes rate of change of x_i, and t' lies between t and $t + \Delta t$. In the simple scheme, $t' = t$, which essentially yields the forward difference integration in time. Here, it is usually necessary to preselect a small value of Δt in order to assure acceptable accuracy.

The foregoing scheme has been modified by Desai[62] to include an iterative procedure in which alternative locations for computing $\dot{x}_i$ can be used. Based on the Lipschitz's condition[80] the size of Δt is increased or decreased automatically such that convergence and stability are assured at each time step.

Neuman and Witherspoon[29] formulated the problem by using the variational principle, Equation 8.10. The resulting set of non-linear differential equations was integrated in time by using Crank–Nicolson procedure. From quantitative results, this scheme was considered to be unconditionally stable.

McCorquodale[28] used the concept of Lagrangian coordinates for flow governed by a non-Darcy law, Equation 8.2. Use of the variational principle, Equation 8.11, yielded a set of non-linear equations that were solved by using the successive over relaxation (SOR) method. The stability of the procedure was found to depend on the size of Δt.

Cheng and Li[41] located the free surface by solving the following form of the equation governing the free surface and by using forward difference in time

$$\left(1 - \frac{\partial \phi}{\partial y}\right)\left(\frac{\partial h}{\partial x}\right)^2 - k_r \frac{\partial \phi}{\partial y} = \frac{\partial h}{\partial t}$$ (8.13)

where $k_r = k_y/k_x$.

8.5.2   Unsaturated flow

The concept of free surface is valid for many practical problems such as flow through silty, sandy and granular media. For certain situations, it is useful to look at the problems as saturated–unsaturated flow and aim at determination of the zone of separation of the two regimes rather than as distinct free surface.

Unsaturated or partially saturated problems using the FD method have been solved for the free surface by satisfying the conditions of zero pressure, Reisenauer,[73] Taylor and Luthin.[81] Desai[27,40,75] used a procedure in which the fluid heads were computed in the entire flow domain and the free surface was located by finding those points where the total head equalled their elevation heads. The exit point on the surface of seepage was located by an iterative procedure coupled with the method of fragments. In Chapter 10, Neuman has proposed a general procedure, in which the so-called free surface is located by finding points where the pressure head vanishes and the surface of seepage is handled by a special iterative method.

The advantage of these approaches is that we can avoid the necessity of collapsing the FE mesh as required in the foregoing schemes, Section 8.5.1. Alternative procedures based on the concept of space-time FE complete discretization with the Galerkin's method[36,82–84] have been proposed. General use of this concept will require further studies toward its evaluation *vis-à-vis* the conventional semidiscretization procedures.

8.6   Flow through discontinuous media

Many applications of the FE method have assumed that the discontinuous or jointed (geologic-rock) medium is incompressible and that the flow occurs only through the joints. Darcy's law has often been assumed;[44,46] however, it has been observed that the flow can be non-linear even at low hydraulic gradients. A non-linear law based on laboratory experiments on flow through fractures reported by Maini, Noorishad and Sharp[85] is

$$C\frac{\partial p}{\partial r} = (v_{\mathrm{m}})^n \tag{8.14}$$

where v_{m} = mean velocity, n shows degree of non-linearity, $\partial p/\partial r$ = hydraulic gradient, and C = constant dependent on the geometry of fractures and viscosity of fluid. Wittke, Rissler and Semprich[86] have reviewed theories of laminar and turbulent flows in open and filled joints or fissures.

For laminar flow through a joint assumed to have parallel sides, the velocity components can be derived from the Navier–Stokes equations as[27,33,44,64,86]

$$v_{xj} = \frac{-g(2a_j)^2}{12v}\frac{\partial \phi}{\partial x} = -k_j\frac{\partial \phi}{\partial x} \tag{8.15a}$$

$$v_{yj} = \frac{-g(2a_j)^2}{12v}\frac{\partial\phi}{\partial y} = -k_j\frac{\partial\phi}{\partial y} \qquad (8.15b)$$

where a_j = half width of joint and v = kinematic viscosity.

A number of schemes are used for simulating flow through fissures. An open joint can be treated as a discrete conduit with permeabilities computed as in Equation 8.15 and porosity equal to unity.[76] Special joint elements for open and filled joints can be used with their permeability matrices established on the basis of the concept similar to that used by Goodman and coworkers,[87] Zienkiewicz and coworkers[88] and Ghaboussi and coworkers.[89]

A triangular FE was used for joints with variable aperture in which the interference due to intersecting fissures was included.[44] In case of a large number of joints, the network of planar conduits was replaced by a homogeneous anisotropic continuous domain with the permeability evaluated on the basis of the geometry of the joints.[86]

Investigations towards numerical solutions for flow through jointed media have been of relatively recent origin and are expected to experience increased research activities. The questions of obtaining realistic numerical models and of measuring precise values of material properties will be of vital importance. The latter will require sophisticated laboratory and field techniques. Moreover, the concept of jointed media replaced by statistically equivalent continuous media will need investigations.

8.7 Factors affecting applications

8.7.1 Numerical characteristics

The practitioner often evaluates suitability of a numerical procedure by solving a number of problems. This quantitative and pragmatic approach is common and often necessary, but it may not necessarily yield a general solution procedure. One way to establish generality of a scheme is to study its numerical characteristics such as convergence, accuracy, stability and consistency. Studies for the numerical properties of the FD procedures are relatively better established in comparison with those of the FE procedures.

From the user's viewpoint, knowledge of the criteria for convergence and stability is desirable to judge the suitability of a given procedure and to arrive at optimum spatial and temporal mesh layouts. We shall state some of these works for both FE and FD procedures together with works that are relevant to the flow problem.

Criteria based on mathematical derivations for various FD schemes have been reported in a number of textbooks and publications.[2,90–95] Convergence and stability criteria for the FD procedures based on quantitative analyses have been presented by various investigators.[1,2,73]

Criteria based on quantitative exercises for the FE procedures have been obtained.[22,28,32,40,96,97] A few studies have discussed and examined FE procedures relevant to the flow problem, and have derived criteria in closed forms.[82,98–102] A procedure based on the Lipschitz condition in which convergence and stability are assured at each time level is proposed by Sandhu, Rai and Desai.[62]

The question of mathematical properties of various FE procedures for problems governed by non-linear equations and boundary conditions is important and will require significant future effort. Moreover, it is necessary to evaluate various numerical schemes and subschemes from viewpoints of the numerical characteristics, computational times and other significant factors.[26,40,83,97,99,104–106] A user is not expected to know details of the vast number of numerical procedures that are available; such evaluations can help in selecting the most economical and suitable scheme for his specific needs.

8.7.2   Order of approximation models

Three finite elements, 4-, 8- and 12-node isoprametric, were used in the FE procedure, Equation 8.12. Computed heads at the free surface given by the three formulations were compared with observed heads in a viscous flow model.[1,27] This study showed that all the three numerical results yielded equally acceptable accuracy. One of the reasons may be that the higher orders, although they improve distribution of potentials within the element and compatibility of nodal potentials, do not seem to improve computation of velocities. Perhaps use of an element with complete polynomials as approximating models may improve the accuracy. However, the computational time increased with the order of the models; for fixed number of nodes, the approximate ratio of time was found to be $1:6:10$.[40] The 8-node element with quadratic variation of ϕ may yield optimum accuracy and economy. The study of Reference 40, however, shows that if a reasonably fine mesh is employed, the simple 4-node element can be more suitable. The subject of trade-offs between accuracy and economy in use of higher order models for the general flow problem is important and will require additional work for definitive conclusions.

8.7.3   Discretization of 'infinite' media

In the event that such natural features as an impervious base (rock), Figure 8.1, are available, the discretized zones can be defined easily. For 'infinite' flow domains such as long river banks and tidal beaches, it is required to establish the significant zones to be included in the FE discretization. This will involve determination of the geometrical extents of the zones and of adequate potential and/or flow conditions on the discretized boundaries.

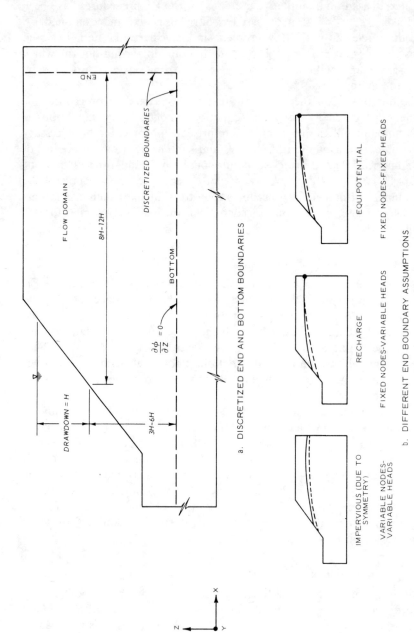

Figure 8.2 Discretization of 'infinite' media (from References 33, 38)

Criteria for evaluation of significant extents and for boundary conditions are often obtained on the basis of quantitative exercises. Different extents of discretized zones and different possible boundary conditions are chosen and corresponding numerical solutions are obtained. Their influence is assessed by comparisons with closed form and/or experimental solutions. Such quantitative analyses for transient unconfined flow (in river banks) have been reported by Desai.[33,35] The numerical solutions were compared with experimental results from a large parallel plate viscous flow model. Figure 8.2(a) shows the significant extents in terms of total drawdown *H*. It was found that the lateral boundary can be fixed at a distance of about 8*H* to 12*H* from the final point of drawdown and an impervious base can be assumed at a distance of about 3*H* to 6*H* from that point. It was also found that assumption of different boundary conditions (Figure 8.2(b)) can influence the numerical solutions. Their choice will depend upon the geological properties of the medium near the discretized boundaries. Criteria for boundary conditions in steady unconfined flow in embankment for tailing ponds have been discussed by Kealy and Williams,[20] and for transient unconfined flow in tidal beaches by Fang and Wang.[37]

8.8 Examples of application

The FE and FD methods have been used for solution of a large number of problems; we shall consider herein only a few applications.

8.8.1 Transient unconfined flow through earth banks[38]

Figure 8.3(a) shows a typical cross-section along the Mississippi River and Figure 8.3(b) shows material properties, locations of piezometers in the bank and the history of river level and corresponding heads measured by the piezometers. The FE mesh used is shown in Figure 8.4(a). Permeability values were obtained on the basis of laboratory measurements and an average value of 40 ft/day was adopted. Values of porosity equal to 0·4 and $\Delta t = 0.025$ day were used.

The FE procedure described previously, Equation 8.12, was employed. Comparisons between predicted heads in the two piezometers and the observations are shown on Figure 8.4(b).

The investigations were aimed at practical (design-analysis) applications, hence, comprehensive analyses for studying the influences of such parameters as permeabilities, geometry of slopes, mesh layouts, river history (rate of drawdown), location of base layers, and finite elements of various orders (4-, 8- and 12-node isoparametric) were performed.[40] Design curves for some common geometries and material properties occurring in the river banks were also generated.

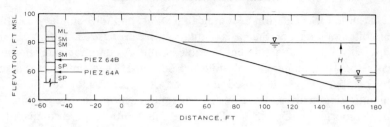

a. CROSS SECTION AT KING'S POINT REVETMENT

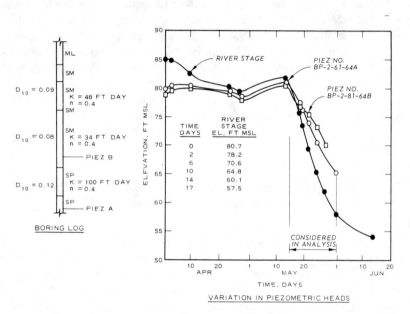

VARIATION IN PIEZOMETRIC HEADS

b. HISTORY OF HEADS AND SOIL PROPERTIES KING'S POINT REVETMENT

Figure 8.3 Details of river bank, soil properties and observed history of heads (from Reference 38)

The quantitative analyses showed the following stability criterion for the transient free surface problem:

$$\bar{k}\,\Delta t\left(\frac{1}{\Delta \bar{x}^2} + \frac{1}{\Delta y^2}\right) R \leqslant 0\cdot 05 \qquad (8.16)$$

where $\bar{k}$ = permeability = $(k_x^2 + k_y^2)^{\frac{1}{2}}$, $\Delta \bar{x}, \Delta \bar{y}$ = (mean) element dimensions, R = rate of drawdown, and Δt = time increment.

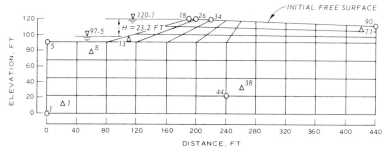

a. FINITE ELEMENT MESH

$k_x = k_y = 40$ FT DAY, $\Delta t = 0.025$ DAYS, $n = 0.4$

TIME DURING DRAWDOWN DAYS	PIEZOMETER A		PIEZOMETER B	
	MEASURED FT	PREDICTED FT	MEASURED FT	PREDICTED FT
0	80.7	79.0	79.5	79.3
2	79.0	76.9	77.8	77.1
6	74.7	72.3	75.2	72.4
10	70.2	68.8		
14	66.9	66.1		
17	65.0	64.6		

b. COMPARISON BETWEEN PREDICTIONS AND FIELD DATA

Figure 8.4 Finite element mesh and comparisons between computations and observations (from Reference 38)

8.8.2 Non-Darcy flow through rock-fill dam

We shall consider an example from the work of McCorquodale[28] that involves unsteady unconfined flow through a laboratory model of rock-fill dam, Figure 8.5(a). The formulation was based on the governing Equation 8.6 and non-Darcy flow rule, Equation 8.2a. The variational functional, Equation 8.11, was expressed as

$$\Pi(\phi) = \int_t^{t+\Delta t} \int\int_{\Omega(t)} \left[k(|\nabla\phi|) \left[\left(\frac{\partial\phi}{\partial x}\right)^2 + \left(\frac{\partial\phi}{\partial y}\right)^2 \right] \right. $$

$$\left. + \frac{a}{6bc'}(2 - c'|\nabla\phi|)\sqrt{1 + c'|\nabla\phi|} \right] dx\, dy\, dt \qquad (8.17)$$

Here $\phi\, (x, y, 0)$ are known.

The SOR[102] method was used to solve the non-linear algebraic equations. Locations of the free surface were determined by using the Lagrangian approach

$$\vec{FS}(x, y, t + \Delta t) = \int_t^{t+\Delta t} \vec{v}(x, y, t')\, dt' + \vec{s}(x, y, t) \simeq \frac{\vec{v}\, \Delta t}{n} + \vec{FS}(x, y, 0) \qquad (8.18)$$

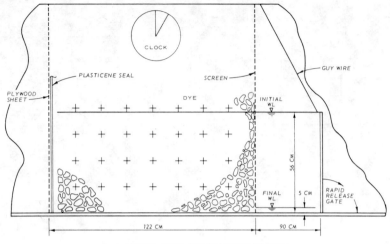

a. TYPICAL MODEL FOR RECTANGULAR BANK

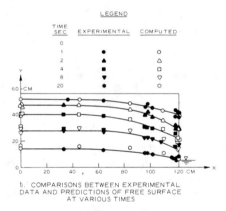

b. COMPARISONS BETWEEN EXPERIMENTAL
DATA AND PREDICTIONS OF FREE SURFACE
AT VARIOUS TIMES

Figure 8.5   Non-Darcy flow through model bank (from Reference 28)

where $FS(x, y, t')$ = location of a fluid particle on the free surface at time t'; x, y = rectangular coordinates of the particle at t'; $\bar{v}$ = average 'bulk' velocity during Δt.

Figure 8.5(b) shows comparisons between the predicted and observed locations of the free surface. On the basis of a quantitative analyses, the stability criteria was reported as

$$\Delta t \leqslant \frac{\Delta \bar{x}}{\sqrt{g y_s}} \tag{8.19}$$

where $\Delta \bar{x}$ = average horizontal spacing of the free surface nodes and y_s = depth to the free surface.

8.8.3 Aquifer analysis and sudden drawdown—comparison between FD and FE methods

Investigations by using both the FD–alternating direction implicit (ADI) procedure and a FE scheme based on the Galerkin's method for analysis of Musquodoboit Harbor Aquifer, Nova Scotia, Canada, were performed by Pinder and Frind.[26] Figure 8.6 shows the cross section of the aquifer and

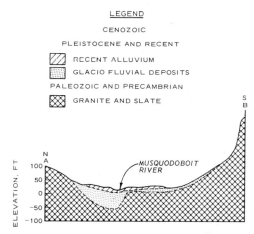

Figure 8.6   Cross-section of Musquodoboit Harbor aquifer (from Reference 26)

Figures 8.7(a) and 8.7(b) illustrate the FD and FE meshes respectively. Values of transmissivities were determined from pumping tests. The FE equations were formulated by using the Galerkin's method in which isoparametric elements of various orders were used. Comparisons between the predictions of drawdown in a well in the infinite aquifer are shown in Figure 8.8.

Desai[38] used an FD–alternating direction explicit (ADE) procedure and the FE method for predicting free surface after sudden drawdown in a parallel plate viscous flow model. The properties of the laboratory model[107] and the FE mesh are shown in Figure 8.9(a) and comparisons of FD and FE and laboratory observations are included in Figure 8.9(b).

On the basis of the conclusions by Pinder and Frind and by Desai,[38,40] the following comments are offered concerning the comparison of the FD and FE methods for the problem of seepage. Both methods possess advantages and their choice will depend upon the particular problem on hand. For instance, for transient seepage in homogeneous media with relatively regular boundaries, the FD method will be more economical and equally acceptable from the viewpoint of accuracy. The FE method, with carefully designed meshes and higher order of isoparametric elements can yield the same accuracy for lesser number of equations as compared with the FD

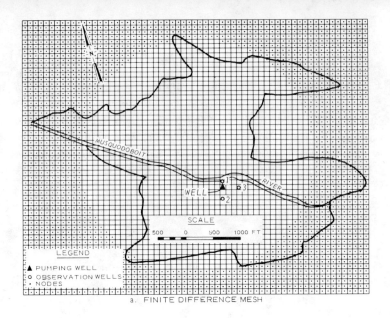

a. FINITE DIFFERENCE MESH

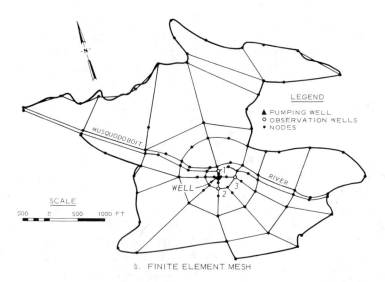

b. FINITE ELEMENT MESH

Figure 8.7 Finite difference and finite element meshes for aquifer (from Reference 26)

method. With efficient solution schemes, the computational effort in the FE method can be competitive with the efficient FD-ADI procedures.

The formulations and development aspects of FE procedures can be more complex than the FD schemes. Greater care is required for error-free input data for the FE procedures than that for the FD procedures.

Finally, the FE method with the Galerkin's procedure can provide highly general and versatile formulations for a wide range of problems in fluid flow through porous media.

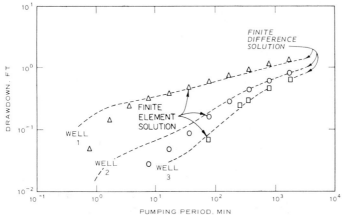

Figure 8.8 Comparisons between finite element and finite difference predictions (from Reference 26)

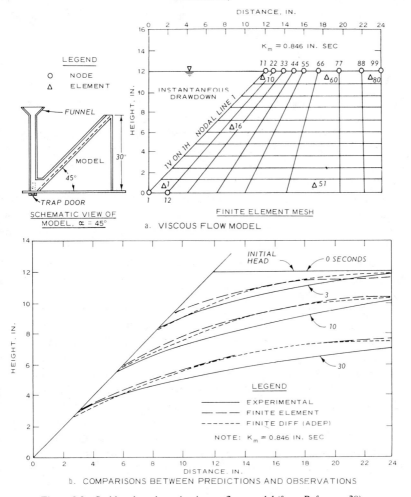

a. VISCOUS FLOW MODEL

b. COMPARISONS BETWEEN PREDICTIONS AND OBSERVATIONS

Figure 8.9 Sudden drawdown in viscous flow model (from Reference 38)

8.9 Some limitations

Computational difficulties can arise for media with abrupt variations in material properties and geometrical configurations. For example, location of free surface at intersection of core and shell in a dam will require special schemes to avoid difficulties.

The formulations covered herein may not be suitable for all problems. For instance, the concept of free surface may not be valid for highly cohesive soils in which significant capillary fringes may exist and the pore spaces can remain saturated under large suction. The assumption of a rigid soil–water system can limit the height of zone of flow that can be included. The rate of drawdown can have significant influence on the numerical scheme; the rate may have to be limited in order that the pressure lag caused by the compressibility of the system is small so as to justify the assumptions.[1,32,40] Inclusion of inertia effects will require further investigations.

8.10 Comments and conclusions

Fluid flow through porous media is one of the subjects for which both the FE and FD methods have been used extensively and found to be highly successful. Both possess advantages for specific problems and in some instances can yield similar formulations and solutions. As compared with the FD method, the FE method can provide general formulations that have applications for a wider range of problems.

Most of the FE developments used in the past have used the fluid potential as the unknown with corresponding variational functionals. In some cases, where the continuity of flow (across interfaces in layered media) is important and more accurate evaluation of quantities of seepage is desired, formulations based on the stream function or velocities as the unknown may be more appropriate and relevant. Use of the mixed formulations can also be suitable for certain situations.

The flow is generally non-linear over the laminar regime and Darcy's law can be valid only for a portion of the laminar regime at small velocities. Further work will be required for non-linear flow and flow in the turbulent regime. It may be mentioned, however, that Darcy's law for non-linear flow may often not yield precise values of the quantity of flow, but can yield satisfactory evaluation of distribution or potentials and flow nets. Thus it can be adequate for computation of seepage forces and for stability analyses.[21,40]

Development of adequate constitutive laws seem to have lagged behind the development of numerical methods. Sophisticated laboratory and field techniques are desirable for determination of adequate parameters. This is particularly true for flow through jointed media. This topic is expected to see increased research in the future towards theoretical developments, field techniques for mapping discontinuities, and determination of their constitutive parameters.

Although the approach for verifying a numerical procedure on the basis of quantitative analyses may often be necessary, significant work will be warranted in establishing mathematical properties of various numerical procedures for problems governed by non-linear equations and boundary conditions. The question of numerical stability has not been answered satisfactorily. Additional research will be needed to establish general criteria and procedures. Moreover, selection of optimum meshes, spatial and temporal, may not necessarily be governed only by mathematical convergence and stability. Required accuracy can be a significant factor for the time-dependent problem.[84,96,97,99] It is also necessary to evaluate the large number of available procedures from the viewpoints of accuracy, reliability and economy; this will provide criteria that can assist the user in selecting optimum schemes. The analyst will have to devote attention to this pursuit in order that the application of the fast growing variety of techniques does not lag much behind their use for solution of real-life problems.

On the basis of the correlations between numerical solutions and observations that have been obtained already, the numerical methods can be advocated as efficient tools for design-analysis. In order to establish confidence in the procedures as means for (final) field designs, however, the additional work and correlations will be required.

References

1. C. S. Desai, 'Seepage in porous media', in *Numerical Methods in Geotechnical Engineering*, C. S. Desai and J. T. Christian (eds.), McGraw-Hill, New York, (forthcoming).
2. I. Remson, G. M. Hornberger and F. J. Molz, *Numerical Methods in Subsurface Hydrology*, Wiley-Interscience, New York, 1971.
3. O. C. Zienkiewicz, P. Mayer and Y. K. Cheung, 'Solution of anisotropic seepage by finite elements', *Journal of Engineering Mechanics Division, ASCE*, **92**, no. EM1 (1966).
4. O. C. Zienkiewicz and C. J. Parekh, 'Transient field problems: two-dimensional and three-dimensional analysis by isoparametric finite element', *International Journal of Numerical Methods in Engineering*, **2**, 1 (1970).
5. O. C. Zienkiewicz and D. J. Naylor, 'Finite element studies of soils and porous media', *Lectures on Finite Element Methods in Continuum Mechanics*, J. T. Oden and E. R. A. Oliveira (eds.), UAH Press, Huntsville, 1973, pp. 459–493.
6. O. C. Zienkiewicz, *The Finite Element Method in Engineering Science*, McGraw-Hill, London, 1971.
7. C. S. Desai and J. F. Avel, *Introduction to the Finite Element Method; A Numerical Method for Engineering Analysis*, Van Nostrand Reinhold, New York, 1972.
8. C. S. Desai, 'Overview, trends and projections: theory and applications of the finite element method in geotechnical engineering', State-of-the-Art Paper, *Proceedings of Symposium on Application of FEM in Geotechnical Engineering*, C. S. Desai (ed.), USA Engineer Waterways Experiment Station, Vicksburg, 1972.

9. C. S. Desai, 'Finite element procedures for seepage analysis using an isoparametric element', *Proceedings of Symposium on Application of FEM in Geotechnical Engineering*, C. S. Desai (ed.), USA Engineer Waterways Experiment Station, Vicksburg, 1972.

10. C. S. Desai (ed.), *Proceedings of Symposium on Application of FEM in Geotechnical Engineering*, USA Waterways Experiment Station, Vicksburg, 1972.

11. J. C. Cavendish, H. S. Price and R. S. Varga, 'Galerkin methods for the numerical solution of boundary value problems', *Journal of Society of Petroleum Engineers*, **9**, 2 (1969).

12. 'Finite element solution of steady state potential flow problems', *Hydrologic Engineering Center (HEC)*, *No. 723-G2-L2440*, USA Engineer District, Sacramento, California, Nov. 1970.

13. P. Guellec, 'Calculation of flows in porous media by the FE method', *Rapport de Recherche No. 11*, Laboratories des Ponts et Chaussees, Nov. 1970.

14. G. J. W. King and R. N. Chowdhury, 'Finite element solution for quantity of steady seepage', *Civil Engineering and Public Works Review*, **66**, 785 (1971).

15. R. L. Taylor and C. B. Brown, 'Darcy flow solutions with a free surface', *Journal of Hydraulics Division, ASCE*, **93**, HY2 (1967).

16. W. D. L. Finn, 'Finite element analysis of seepage through dams, *Journal of Soil Mechanics and Foundations Division, ASCE*, **93**, SM6 (1967).

17. J. D. Fenton, 'Hydraulic and stability analyses of rockfill dams', *DR No. 15*, Department of Civil Engineering, University of Melbourne, July 1968.

18. R. E. Volker, 'Nonlinear flow in porous media by finite elements', *Journal of the Hydraulics Division, ASCE*, **95**, HY6 (1969).

19. S. P. Neuman and P. A. Witherspoon, 'Finite element method for analyzing steady seepage with a free surface', *Water Resources Research*, **6**, 3 (1970).

20. C. D. Kealy and R. E. Williams, 'Flow through a tailing pond embankment', *Water Resources Research*, **7**, 4 (1971).

21. D. H. Trollope, K. P. Stark and R. E. Volker, 'Complex flow through porous media', *The Australian Geomechanics Journal*, **G1**, 1 (1971).

22. C. S. Desai, 'Free surface seepage through foundation and berm of cofferdams', *Indian Geotechnical Journal*.

23. I. Javandel and P. A. Witherspoon, 'Application of the finite element method to transient flow in porous media', *Transactions, Society of Petroleum Engineers*, **243**, 241–251 (Sept. 1968).

24. I. Javandel and P. A. Witherspoon, 'A method of analyzing transient fluid flow in multilayered aquifers', *Water Resources Research*, **5**, 4 (1969).

25. S. P. Neuman and P. A. Witherspoon, 'Transient flow of ground water to wells in multiple-aquifer systems', *Report 69-1*, Geotechnical Engineering, University of California at Berkeley, California, Jan. 1969.

26. G. F. Pinder and E. O. Frind, 'Application of Galerkin's procedure to aquifer analysis', *Water Resources Research*, **8**, 1 (1972).

27. C. S. Desai, 'Seepage in Mississippi river banks: analysis of transient seepage using viscous flow model and numerical methods', *Miscellaneous Paper S-70-3*, USA Engineer Waterways Experiment Station, Vicksburg, Feb. 1970.

28. J. A. McCorquodale, 'Variational approach to non-Darcy flow', *Journal of the Hydraulics Division, ASCE*, **96**, HY11 (1970).

29. P. W. France, C. J. Parekh, J. C. Peters and C. Taylor, 'Numerical analysis of free surface seepage problems', *Journal of Irrigation and Drainage Division, ASCE*, **97**, IR1 (1971).

30. S. P. Neuman and P. A. Witherspoon, 'Analysis of unsteady flow with a free surface using the finite element method', *Water Resources Research*, **7**, 3 (1971).

31. W. Harrison, C. S. Fang and S. N. Wang, 'Groundwater flow in a sandy tidal beach 1. One-dimensional finite element analysis', *Water Resources Research*, 7, 5 (1971).
32. L. T. Isaacs and K. G. Mills, Discussion to paper 'Numerical analysis of free surface seepage problems', *Journal of Irrigation and Drainage Division, ASCE*, 98, IR1 (1972).
33. C. S. Desai, 'Seepage analysis of earth banks under drawdown', *Journal of Soil Mechanics and Foundations Division, ASCE*, 98, SMI (1972).
34. J. D. Tulk and G. P. Raymond, 'Drainage of granular soils', *Proceedings, Specialty Conference on SCM in Civil Engineering*, McGill University, Montreal, Canada, June 1972.
35. C. S. Desai, 'An approximate solution for unconfined seepage', *Journal of Irrigation and Drainage Division, ASCE*, 99, IR1 (1973).
36. J. C. Bruch, 'Nonlinear equation of unsteady ground-water flow', *Journal of Hydraulics Division, ASCE*, 99, HY3 (1973).
37. C. S. Fang and S. N. Wang, 'Groundwater flow in a sandy tidal beach 2. Two-dimensional finite element analysis', *Water Resources Research*, 8, 1 (1973).
38. C. S. Desai, 'Seepage in Mississippi river banks, analysis of transient seepage using viscous flow model, and finite difference and finite element methods', *Technical Report 1*, USA Engineer Waterways Experiment Station, Vicksburg, May 1973.
39. Y. H. Huang, 'Unsteady flow toward an artesian well', *Water Resources Research*, 9, 2 (1973).
40. C. S. Desai, Y. S. Jeng and R. S. Sandhu, 'Seepage in Mississippi river banks; analysis and design by numerical procedures and computer codes', *Technical Report 2*, USA Engineer Waterways Experiment Station, Vicksburg, to be published.
41. R. T. Cheng and C. Y. Li, 'On the solution of transient free-surface flow problems in porous media by the finite element method', *Journal of Hydrology*, 20 (1973).
42. W. Wittke, 'Methods of calculation of three-dimensional problems of percolation of fissured rock by finite elements and resistance networks', *Proceedings, 2nd Congress International Society of Rock Mechanics*, Vol. 3, Belgrade, 1970.
43. W. Wittke, 'Three-dimensional percolation in fissured rock', *Proceedings, Symposium on Planning Open-Pit Mines*, Johannesburg, South Africa, 1970.
44. C. R. Wilson and P. A. Witherspoon, 'An investigation of laminar flow in fractured porous media', *Geotechnical Engineering Report No. 70-6*, University of California, Berkeley, 1970.
45. J. Noorishad, P. A. Witherspoon and T. L. Brekke, 'A method for coupled stress and flow analysis of fractured rock masses', *Geotechnical Engineering Report No. 71-6*, University of California, Berkeley, 1971.
46. *Proceedings of the Symposium on Percolation Through Fissured Rock*, Stuttgart, W. Germany, 1972.
47. H. S. Price, J. C. Cavendish and R. S. Varga, 'Numerical methods of higher-order accuracy for diffusion-convection equations', *Journal of Society of Petroleum Engineers*, 8, 3 (1968).
48. G. L. Guymon, 'A finite element solution of the one-dimensional diffusion-convection equation', *Water Resources Research*, 6, 1 (1970).
49. G. L. Guymon, V. H. Scott and L. R. Herman, 'A general numerical solution of the two-dimensional diffusion-convection equation by the finite element method', *Water Resources Research*, 6, 6 (1970).
50. G. L. Guymon, 'Note on the finite element solution of the diffusion-convection equation', *Water Resources Research*, 8, 5 (1972).

180 Finite Elements in Fluids

51. M. Nalluswami, R. A. Longenbaugh and D. K. Sunada, 'Finite element method for the hydrodynamic dispersion equation with mixed partial derivatives', *Water Resources Research*, **8**, 5 (1972).
52. R. S. Sandhu and E. L. Wilson, 'Finite element analysis of seepage in elastic media', *Journal of the Engineering Mechanics Division, ASCE*, **95**, EM3 (1969).
53. J. T. Christian and J. W. Boehmer, 'Plane strain consolidation by finite elements', *Journal of the Soil Mechanics and Foundations Division, ASCE*, **96**, SM4 (1970).
54. Y. Yokoo, K. Yamagata and H. Nagaoka, 'Finite element method applied to Biot's consolidation theory', *Soils and Foundations, Japan Society of Soil Mechanics and Foundation Engineering*, **11**, 1 (1971).
55. Y. Yokoo, K. Yagamata and H. Nagaoka, 'Variational principles for consolidation', *Soils and Foundations*, **11** (1971).
56. C. T. Hwang, N. R. Morgenstern and D. W. Murray, 'On solutions of plane strain consolidation problems by FE methods', *Canadian Geotechnical Journal*, 8 (1971).
57. C. T. Hwang, N. R. Morgenstern and D. W. Murray, 'Application of the FE method to consolidation problems', *Proceedings, Symposium on Application of FEM in Geotechnical Engineering*, C. S. Desai (ed.), USA Engineer Waterways Experiment Station, Vicksburg, 1972.
58. R. S. Sandhu, 'Finite element analysis of consolidation and creep', *Proceedings, Symposium on Application of FEM in Geotechnical Engineering*, C. S. Desai (ed.), USA Engineer Waterways Experiment Station, Vicksburg, 1972.
59. J. Ghaboussi and E. L. Wilson, 'Seismic analysis of earth dam-reservoir systems', *Journal of the Soil Mechanics and Foundations Division, ASCE*, **99**, SM10 (1973).
60. R. W. Lewis and R. W. Garner, 'A FE solution of coupled electro-kinetic and hydrodynamic flow in porous media', *International Journal for Numerical Methods in Engineering*, **5**, 41–44 (1972).
61. R. W. Lewis and C. Humpheson, 'Numerical analysis of electro-osmotic flow in soils', *Journal of the Soil Mechanics and Foundations Division, ASCE*, **99**, SM8 (1973).
62. J. T. Oden, O. C. Zienkiewicz, R. H. Gallagher and C. Taylor (eds.), *Finite Element Methods in Flow Problems*, University of Alabama (Huntsville) Press, 1974.
63. M. K. Hubbert, 'Darcy's law and the field equations of flow of underground fluids', *Journal of Petroleum Tech.*, **8**, 222–239 (1956).
64. M. E. Harr, *Groundwater and Seepage*, McGraw-Hill, New York, 1962.
65. P. H. Fochheimer, 'Wassebewegung durch boden', *Z. Verdt. Ing.*, p. 782 (1901).
66. A. Missbach, *Listy Cukrova*, **55**, 293 (1937).
67. R. Engelund, 'On the laminar and turbulent flow of ground water through homogeneous sand', *Transactions, Danish Academy of Technical Science*, No. 3, 1953.
68. J. Happel and H. Brenner, *Low Reynolds Number Hydrodynamics*, Prentice-Hall, Englewood Cliffs, N.J., 1965.
69. International Association of Hydraulic Research, *Fundamentals of Transport Phenomena in Porous Media, Development in Soil Science 2*, Elsevier, Amsterdam, New York, 1972.
70. A. E. Scheidegger, 'Deterministic and statistical characterization of porous media and computational methods of analysis', *Fundamentals of Transport Phenomena in Porous Media*, Elsevier, Amsterdam, New York, 1972.
71. A. E. Scheidegger and K. H. Liao, 'Thermodynamic analogy of mass transport processes in porous media', *Fundamentals of Transport Phenomena in Porous Media*, Elsevier, Amsterdam, New York, 1972.

72. D. H. Trollope, 'The mechanics of discontinua on clastic mechanics in rock problems', in *Rock Mechanics in Engineering Practice*, K. G. Stagg and O. C. Zienkiewicz (eds.), Wiley, London, 1968.

73. A. E. Reissenauer, 'Methods for solving problems in multi-dimensional partially saturated steady flow in soils', *Journal of Geophysical Research*, **68**, 20 (1963).

74. P. Ya. Polubarinova-Kochina, *Theory of Ground Water Movement*, Princeton University Press, Princeton, N.J., 1962.

75. C. S. Desai and W. C. Sherman, 'Unconfined transient seepage in sloping banks', *Proc. ASCE, J. of the Soil Mech. and Fdn. Div.*, **97**, SM2 (1971).

76. M. E. Gurtin, 'Variational principles for linear initial-value problems', *Quarterly of Applied Mathematics*, **22**, 3 (1964).

77. S. P. Neuman and P. A. Witherspoon, 'Variational principles for confined and unconfined flow of ground water', *Water Resources Research*, **6**, 5 (1970).

78. B. A. Finlayson, *The Method of Weighted Residuals and Variational Principles*, Academic Press, New York, 1972.

79. M. M. Aral, P. G. Mayer and C. V. Smith, 'Finite element Galerkin method solutions to selected elliptic and parabolic differential equations', *Proceedings, 3rd Conference on Matrix Methods of Structural Mechanics, Wright-Patterson AFB, Dayton, Ohio, Nov. 1971*.

80. P. Henrici, *Elements of Numerical Analysis*, Wiley, New York, 1964.

81. G. S. Taylor and J. N. Luthin, 'Computer methods of transient analysis of water table aquifers', *Water Resources Research*, **15** (Feb. 1969).

82. J. T. Oden, *Finite Elements of Nonlinear Continua*, McGraw-Hill, New York, 1972.

83. G. Zyvoloski and J. C. Bruch, 'Finite element weighted residual solution of one-dimensional unsteady and unsaturated flows in porous media', *UCSM-ME-73-4*, Department of Mechanical Engineering, University of California, Santa Barbara, June 1973.

84. C. S. Desai, J. T. Oden and L. D. Johnson, 'Evaluation and analyses of some finite element and finite difference procedures for time-dependent problems', *Miscellaneous Paper*, USA Engineer Waterways Experiment Station, Vicksburg.

85. Y. N. T. Maini, J. Noorishad and J. Sharp, 'Theoretical and field considerations on the determination of *in situ* hydraulic parameters in fractured rock', *Proceedings, Symposium on Percolation Through Fissured Rock*, Stuttgart, 1972.

86. W. Wittke, P. Rissler and S. Semprich, 'Räumliche, laminare und turbulente strömung in klüftigem fels nach zwei verschiedenen rochenmodellen', *Proceedings, Symposium on Percolation Through Fissured Rock*, Stuttgart, 1972.

87. R. E. Goodman, R. L. Taylor and T. Brekke, 'A model for the mechanics of jointed rock', *Journal of the Soil Mechanics and Foundations Division, ASCE*, **94**, SM3 (1968).

88. O. C. Zienkiewicz and coworkers, 'Analysis of non-linear problems in rock mechanics with particular reference to jointed rock systems', *Proceedings, 2nd Congress Society for Rock Mechanics*, Vol. 3, Belgrade, 1970.

89. J. Ghaboussi, E. L. Wilson and J. Isenberg, 'Finite element for rock joints and interfaces', *Journal of the Soil Mechanics and Foundations Division, ASCE*, **99**, SM10 (1973).

90. R. D. Richtmeyer and K. W. Morton, *Difference Methods for Initial-Value Problems*, Interscience, New York, 1967.

91. G. G. O'Brien, M. A. Hyman and S. Kaplan, 'A study of the numerical solution of partial differential equations', *Journal of Mathematics and Physics*, **XXIX**, 3, 223–251 (1951).

92. D. Young, 'The numerical solution of elliptic and parabolic partial-differential equations', *Survey of Numerical Analysis*, J. Todd (ed.), McGraw-Hill, New York, 1962.
93. G. Terzidis, 'Computational schemes for the Boussinesq equation', *Journal of Irrigation and Drainage Division, ASCE*, **94**, IR4 (1968).
94. J. L. Milton and W. P. Goss, 'Stability criteria for explicit finite difference solutions of the parabolic diffusion equation with nonlinear boundary conditions', *International Journal Numerical Methods in Engineering*, **7** (1973).
95. A. M. Clausing, 'Practical techniques for estimating the accuracy of finite-difference solutions to parabolic equations', *Journal of Applied Mechanics*, **40**, Series E, No. 1, (1973).
96. C. S. Desai and L. D. Johnson, 'Evaluation of some numerical schemes for consolidation', *International Journal Numerical Methods in Engineering*, **7** (1973).
97. C. S. Desai and L. D. Johnson, 'Evaluation of two finite element formulations for one-dimensional consolidation', *Computers and Structures*, **2**, 469–486 (1972).
98. G. Strang and G. J. Fix, *An Analysis of the Finite Element Method*, Prentice-Hall, Englewood Cliffs, N.J., 1973.
99. C. S. Desai and R. L. Lytton, 'Stability criteria for two finite element schemes for parabolic equation', to be published.
100. R. S. Varga, *Functional Analysis and Approximation Theory in Numerical Analysis*, Society for Industrial and Applied Mathematics, 1971.
101. J. Douglas and T. Dupont, 'Galerkin methods for parabolic equations', *SIAM Journal of Numerical Analysis*, **7**, 4 (1970).
102. J. Douglas and T. Dupont, 'A finite element collocation method for quasilinear parabolic equations', *Mathematics of Computations*, **27**, 21 (1973).
103. R. V. S. Yalamanchili and S. C. Chu, 'Stability and oscillation characteristics of finite element, finite difference, and weighted residuals methods for transient two-dimensional heat conduction in solids', *Journal of Heat Transfer, Transactions, ASME*, **95**, 2, Series C (1973).
104. A. F. Emery and W. W. Carson, 'An evaluation of the use of the FE method in the computation of temperature', *Journal of Heat Transfer*, **93**, 2, Series C (1971).
105. R. L. Schiffman, 'Efficient use of computer resources', in *Proceedings, Symposium on Applications of FE Method in Geotechnical Engineering*, C. S. Desai (ed.), USA Engineer Waterways Experiment Station, Vicksburg, 1972.
106. J. C. Bruch and G. Zyvoloski, 'A finite element weighted residual solution to one-dimensional field problems', *International Journal Numerical Methods in Engineering*, **6**, 577–585 (1973).
107. A. H. Dvinoff and M. E. Harr, 'Phreatic surface location after drawdown', *Journal of the Soil Mechanics and Foundation Division, ASCE*, **97**, SM1 (1971).

Chapter 9

A Finite Element Approach to Two-Phase Flow in Porous Media

R. W. Lewis, E. A. Verner and O. C. Zienkiewicz

9.1 Introduction

The displacement of oil from the pores of a hydrocarbon bearing medium by the injection of an immiscible fluid has been studied extensively in the oil industry. One of the most versatile methods of predicting this displacement performance is by the numerical solution of a mathematical model. The numerical solution has often been obtained by the method of finite-differences but the Galerkin approach[1] and the method of characteristics[2] have also been used. This paper presents a technique based on the finite element variant of the Galerkin method and describes a computer program which includes linear, parabolic and cubic isoparametric elements.

The application of the finite element method to spatial approximation results in a set of simultaneous, non-linear algebraic-differential equations. The method chosen for the solution of these equations involves a family of single step methods[3] which depend on a parameter γ. By varying the value of γ, the scheme yields either the Euler or forward-difference scheme, the Crank–Nicolson or trapezoidal rule, or the backward Euler or backward-difference scheme.

A one-dimensional axisymmetric problem and a two-dimensional water/oil coning problem are solved to illustrate the versatility of the program. These are compared with solutions obtained by other numerical techniques and are found to compare favourably in all cases. The program is general in that any arbitrary shape of flow domain can be analysed along with various forms of boundary conditions. Also, no restrictions are imposed regarding the homogeneity and isotropy of the porous medium.

9.2 The two-phase mathematical model

The space occupied by the porous medium is referred to as the region Ω with boundary Γ. The subscripts 'o' and 'w' refer to the oil and water phases, respectively.

The flow of oil (v_o) per unit area across the direction of flow is represented by the relationship

$$\mathbf{v}_o = -\frac{1}{\mu_o}(\mathbf{k}\,\mathbf{k}_{ro})(\nabla p_o + \rho_o g\,\nabla h) \qquad (9.1a)$$

where $(\mathbf{k}\,\mathbf{k}_{ro})$ is the permeability matrix whose terms are given individually by the product $k_{ij}k_{ro}$, with k_{ij} the local permeability and k_{ro} the relative permeability to oil, μ_o is the oil viscosity, p_o the pressure in the oil phase, ρ_o the oil density, g the acceleration of gravity and h is the height above some datum. In the two-dimensional case the vector $\mathbf{v}_o$ contains the components v_{o_x} and v_{o_y} and ∇p_o and consists of $\partial p_o/\partial x$ and $\partial p_o/\partial y$. Thus, the symbol ∇ represents the vector $L\,\partial/\partial x\,\partial/\partial y J^T$.

Similarly, the flow of water is given by

$$\mathbf{v}_w = \frac{1}{\mu_w}(\mathbf{k}\,\mathbf{k}_{rw})(\nabla p_w + \rho_w g\,\nabla h) \qquad (9.1b)$$

To simplify these expressions we introduce the potentials ψ_o and ψ_w defined as follows

$$\psi_o = p_o + \rho_o g h \qquad (9.2a)$$

$$\psi_w = p_w + \rho_w g h \qquad (9.2b)$$

Also, we introduce the 'mobility matrices' $\mathbf{M}_o$ and $\mathbf{M}_w$

$$\mathbf{M}_o = \frac{1}{\mu_o}(\mathbf{k}\,\mathbf{k}_{ro}) \qquad (9.3a)$$

$$\mathbf{M}_w = \frac{1}{\mu_w}(\mathbf{k}\,\mathbf{k}_{rw}) \qquad (9.3b)$$

Thus, Equations 9.1 and 9.2 become

$$\mathbf{v}_o = -\mathbf{M}_o\,\nabla\psi_o \qquad (9.4a)$$

$$\mathbf{v}_w = -\mathbf{M}_w\,\nabla\psi_w \qquad (9.4b)$$

From continuity considerations the rate of accumulation of each of the fluid phases is given by the product of the porosity v and the rate of increase of saturation, $\dot{S}$. (The dot denotes the derivative of the saturation S with respect to time.) Also, the rate of outflow is the divergence of the velocity. Thus, for each phase

$$\operatorname{div}\mathbf{v}_o + v\dot{S}_o = q_o \qquad (9.5a)$$

$$\operatorname{div}\mathbf{v}_w + v\dot{S}_w = q_w \qquad (9.5b)$$

where

$$\text{div } \mathbf{v}_o = \frac{\partial v_o}{\partial x} + \frac{\partial v_o}{\partial y} \tag{9.6a}$$

$$\text{div } \mathbf{v}_w = \frac{\partial v_w}{\partial x} + \frac{\partial v_w}{\partial y} \tag{9.6b}$$

and q_o and q_w are the total volumetric influxes of oil and water respectively.

The oil and water phase pressures at a point within the medium differ by the capillary pressure p_c, i.e.

$$p_c = p_o - p_w \tag{9.7}$$

It is assumed that capillary forces will cause a displacement of fluid resulting in a certain saturation which depends only on the value of the capillary pressure. An example of such a capillary pressure-saturation curve along with the relative permeability curves is shown in Figure 9.1. From these it is seen that the relative permeability is a *non-linear* function of S_w.

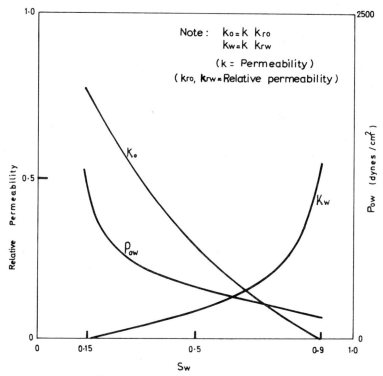

Figure 9.1 Saturation dependent properties

Since the summation of saturations in both phases equals unity, i.e. $S_o + S_w = 1$, only one saturation need be defined. We choose this to be S_w and denote it simply as S. Then, $S_o = 1 - S$ and

$$\dot{S}_o = -\dot{S} \tag{9.8a}$$

$$\dot{S}_w = \dot{S} \tag{9.8b}$$

Also, since the capillary pressure p_c is uniquely related to the saturation we can write

$$\dot{S} = \frac{dS}{dp_c}\frac{\partial p_c}{\partial t} = S'\dot{p}_c \tag{9.9}$$

From Equations 9.2a, 9.2b and 9.7, we can express the capillary pressure as

$$p_c = p_o - p_w = \psi_o - \psi_w + (\rho_w - \rho_o)gh \tag{9.10}$$

and, since $(\rho_w - \rho_o)gh$ is independent of time

$$\dot{p}_c = \dot{\psi}_o - \dot{\psi}_w \tag{9.11}$$

so that we can write Equation 9.9 as

$$\dot{S} = S'(\dot{\psi}_o - \dot{\psi}_w)$$

By use of Equations 9.9 and 9.11, Equation 9.5 becomes

$$\text{div}\,(\mathbf{M}_o\,\nabla\psi_o) = -vS'(\dot{\psi}_o - \dot{\psi}_w) \tag{9.12a}$$

$$\text{div}\,(\mathbf{M}_w\,\nabla\psi_w) = vS'(\dot{\psi}_o - \dot{\psi}_w) \tag{9.12b}$$

Because $\mathbf{M}_o$ and $\mathbf{M}_w$ are matrices whose forms are the relative permeability, which are non-linear functions of the saturation and hence at ψ_o and ψ_w respectively, these are non-linear differential equations.

In a practical analysis situation the potentials ψ_o and ψ_w may be nearly equal, causing difficulties of round-off error in calculation of the capillary pressure from Equation 9.10. Thus, we introduce in place of ψ_o and ψ_w the sum $\psi_s = \psi_o + \psi_w$ and the difference $\psi_d = \psi_o - \psi_w$ of these quantities. By virtue of these substitutions the governing differential equations become

$$\text{div}\,(\mathbf{M}_o + \mathbf{M}_w)\,\nabla\left(\frac{\psi_o + \psi_w}{2}\right) + \text{div}\,(\mathbf{M}_o - \mathbf{M}_w)\,\nabla\left(\frac{\psi_o - \psi_w}{2}\right) = 0 \tag{9.13a}$$

and

$$\text{div}\,(\mathbf{M}_o - \mathbf{M}_w)\,\nabla\left(\frac{\psi_o + \psi_w}{2}\right) + \text{div}\,(\mathbf{M}_o + \mathbf{M}_w)\,\nabla\left(\frac{\psi_o - \psi_w}{2}\right)$$
$$= -2vS'(\dot{\psi}_o - \dot{\psi}_w) \tag{9.13b}$$

With respect to boundary conditions, the surface Γ may be subjected to either a specified flow or potential. For a specified potential $\hat{\Gamma}$ on a portion of the boundary, denoted as Γ_1, we have

$$\psi = \hat{\psi} \quad \text{on } \Gamma_1 \tag{9.14a}$$

and, for a specified flow

$$\hat{V} = (n)^T V \quad \text{on } \Gamma_2 \tag{9.14b}$$

where n is the unit outward normal vector.

For the governing equations as expressed by Equations 9.13 the boundary conditions could involve the sum or difference of the flows or potentials. If the difference in potentials is specified then the saturation level at that boundary is also fixed. Alternatively, it is sometimes convenient to assume that the oil/water ratio at a boundary is dependent on the ratio of the mobilities only. This is expressed by the following condition.

$$\hat{V}_d = \frac{M_d}{M_s} V_s \quad \text{on } \Gamma_3 \tag{9.14c}$$

where M_d and M_s are the difference and the sums of the isotropic mobilities or averages of the principal mobilities. This assumption results in a non-linear boundary condition.

9.3 Formulation of the finite element equations

The finite element representation is derived by use of the method of weighted residuals with Galerkin criterion, i.e. the coefficients of the trial functions are used as weighting factors. Hence, in accordance with the procedures outlined in Chapter 2 and elsewhere in this book, we choose expansions for the potentials ψ_s and ψ_d in terms of shape functions $\mathbf{N}$

$$\psi_s = \mathbf{N}^T \boldsymbol{\psi}_s \tag{9.15a}$$

$$\psi_d = \mathbf{N}^T \boldsymbol{\psi}_d \tag{9.15b}$$

where the row vector $\mathbf{N}^T$ lists the shape functions and the vectors $\boldsymbol{\psi}_s$ and $\boldsymbol{\psi}_d$ list the values of these potentials at the node point of the element.

Application of the weighted residual procedure then yields the following set of algebraic equations

$$\tfrac{1}{2}\mathbf{H}_s\boldsymbol{\psi}_s + \tfrac{1}{2}\mathbf{H}_d\boldsymbol{\psi}_d = \hat{Q}_s \tag{9.16a}$$

$$\tfrac{1}{2}\mathbf{H}_d\boldsymbol{\psi}_s + \tfrac{1}{2}\mathbf{H}_s\boldsymbol{\psi}_d + 2\mathbf{C}\dot{\boldsymbol{\psi}}_d = \hat{Q}_d \tag{9.16b}$$

where

$$\mathbf{H} = \int_\Omega \nabla\mathbf{N}^T M \,\nabla\mathbf{N}\, d\Omega \tag{9.16c}$$

$$\mathbf{C} = \int_\Omega vS'\mathbf{N}^\mathrm{T}\mathbf{N}\,d\Omega \qquad (9.16d)$$

and

$$\mathbf{Q} = \int_\Omega \mathbf{N}^\mathrm{T}\hat{q}\,d\Omega + \int_{\Gamma_2} \mathbf{N}^\mathrm{T}\hat{V}\,d\Gamma \qquad (9.16e)$$

The equations involving $\hat{Q}_\mathrm{s}$ do not contain $\dot{\psi}_\mathrm{s}$ and the system may therefore be described as a set of mixed algebraic-differential equations.

As we have already noted, since the coefficient matrices are functions of capillary pressure (or ψ_d), the Galerkin equations are non-linear in ψ_d and it is preferable to work in terms of ψ_s and ψ_d because for small values of capillary pressure ψ_w and ψ_o may be nearly equal thus giving rise to possible round-off error in the important quantity ψ_d.

9.4   Time integration scheme

The integration of Equations 9.16 with respect to time is carried out by a time-stepping scheme[3] which involves a free parameter γ. This allows variation from the Euler method ($\gamma = 0$), the trapezoidal rule ($\gamma = 0.5$), to the backward difference method ($\gamma = 1$). It is assumed that at time t_{n-1} the unknowns $\boldsymbol{\psi}_\mathrm{d}^{(n-1)}, \dot{\boldsymbol{\psi}}_\mathrm{d}^{(n-1)}$ and $\boldsymbol{\psi}_\mathrm{s}^{(n-1)}$ are known from either a previous solution or are given by the initial conditions. It is then required that at time

$$t_n = t_{n-1} + \Delta t,$$

$$\boldsymbol{\psi}_\mathrm{d}^{(n)} = \boldsymbol{\psi}_\mathrm{d}^{(n-1)} + \Delta t(\gamma\dot{\boldsymbol{\psi}}_\mathrm{d}^{(n)} + (1-\gamma)\dot{\boldsymbol{\psi}}_\mathrm{d}^{(n-1)}) \qquad (9.17)$$

and that Equation 9.16 be satisfied. Liniger and Willoughby[3] discuss methods of choosing suitable values of γ for 'stiff' sets of differential equations (i.e. equations with a wide range of eigenvalues). For the case of $\gamma = 0$, Equations 9.16 and 9.17 uncouple so that two sets of *linear* equations each with half as many unknowns are solved at every time step.

For the general case of $\gamma \neq 0, \boldsymbol{\psi}_\mathrm{d}^{(n)}$ cannot be found directly from Equation 9.17 and the solution must be found using an iterative procedure. The Newton–Raphson method was used to solve these equations which results in the following residuals R for the ith approximation at time t_n:

$$R_\mathrm{s}^{(n,i)} = \tfrac{1}{2}\mathbf{H}_\mathrm{s}\boldsymbol{\psi}_\mathrm{s}^{(n,i)} + \tfrac{1}{2}\mathbf{H}_\mathrm{d}\boldsymbol{\psi}_\mathrm{d}^{(n,i)} - \hat{Q}_\mathrm{s}^{(n,i)} \qquad (9.18a)$$

$$R_\mathrm{d}^{(n,i)} = \tfrac{1}{2}\mathbf{H}_\mathrm{d}\boldsymbol{\psi}_\mathrm{s}^{(n,i)} + \tfrac{1}{2}\mathbf{H}_\mathrm{s}\boldsymbol{\psi}_\mathrm{d}^{(n,i)} + 2\mathbf{C}\dot{\boldsymbol{\psi}}_\mathrm{d}^{(n,i)} - \hat{Q}_\mathrm{d} \qquad (9.18b)$$

$$R_\psi^{(n,i)} = \boldsymbol{\psi}_\mathrm{d}^{(n,i)} - \boldsymbol{\psi}_\mathrm{d}^{(n-1)} - \Delta t(\gamma\dot{\boldsymbol{\psi}}_\mathrm{d}^{(n,i)} + (1-\gamma)\dot{\boldsymbol{\psi}}_\mathrm{d}^{(n-1)}) \qquad (9.18c)$$

The rates of change of the residuals with respect to the unknowns are then used as a means of estimating the required changes in the unknowns to reduce the residuals to zero. These rates of change are given by the coefficient matrices obtained through neglecting higher-order terms in the relation between changes in the residuals and the unknowns. Thus, applying Newton–Raphson to Equations 9.18 yields the following:

$$\Delta R_s^{(n,i)} = \tfrac{1}{2}\mathbf{H}_s\,\Delta\boldsymbol{\psi}_s^{(n,i)} + \tfrac{1}{2}(\mathbf{H}_d + 2\mathbf{E}_s)\,\Delta\boldsymbol{\psi}_d^{(n,i)} \tag{9.19a}$$

$$\Delta R_d^{(n,i)} = \tfrac{1}{2}\mathbf{H}_d\,\Delta\boldsymbol{\psi}_s^{(n,i)} + \tfrac{1}{2}(\mathbf{H}_s + 2\mathbf{G} + 2\mathbf{E}_d + 4\mathbf{B})\,\Delta\boldsymbol{\psi}_d^{(n,i)} + 2\mathbf{C}\,\Delta\dot{\boldsymbol{\psi}}_d \tag{9.19b}$$

$$\Delta R_\psi^{(n,i)} = \Delta\dot{\boldsymbol{\psi}}_d^{(n,i)} - \gamma\,\Delta t\,\Delta\boldsymbol{\psi}_d^{(n,i)} \tag{9.19c}$$

where the matrix $\mathbf{B}$ arises from a change in the matrix $\mathbf{C}$ such that

$$\Delta\mathbf{C}\dot{\boldsymbol{\psi}}_d = \mathbf{B}\,\Delta\boldsymbol{\psi}_d = \left(\int_\Omega vS''\,\dot{\boldsymbol{\psi}}_d\mathbf{N}^T\mathbf{N}\,d\Omega\right)\Delta\boldsymbol{\psi}_d \tag{9.20}$$

and the matrix $\mathbf{E}$ arises from a change in the matrix $\mathbf{H}$ such that

$$\Delta\mathbf{H}\boldsymbol{\psi} = \mathbf{E}\,\Delta\boldsymbol{\psi}_d = \left(\int_\Omega \nabla\mathbf{N}M'\nabla\boldsymbol{\psi}\mathbf{N}\,d\Omega\right)\Delta\boldsymbol{\psi}_d \tag{9.21}$$

Finally, the matrix $\mathbf{G}$ arises from a change in $\hat{Q}_d$ due to a change in mobility ratio for a mobility boundary condition.

$$\Delta\hat{Q}_d = \mathbf{G}\,\Delta\boldsymbol{\psi}_d = \left(\int_{\Gamma_3} \frac{M_d}{M_s}V_s\mathbf{N}^T\mathbf{N}\,d\Gamma\right)\Delta\boldsymbol{\psi}_d \tag{9.22}$$

Equation 9.19c may be used to eliminate $\Delta\boldsymbol{\psi}_d^{(n,i)}$ from Equations 9.19a and 9.19b resulting in the following expression:

$$\Delta R_s^{(n,i)} = \tfrac{1}{2}\mathbf{H}_s\,\Delta\boldsymbol{\psi}_s^{(n,i)} + \tfrac{1}{2}(\mathbf{H}_d + 2\mathbf{E}_s)\,\Delta\boldsymbol{\psi}_d^{(n,i)} \tag{9.23a}$$

$$\left(\Delta R_d + \frac{2}{\gamma\,\Delta t}\mathbf{C}R_\psi\right)^{(n,i)} = \tfrac{1}{2}\mathbf{H}_d\,\Delta\boldsymbol{\psi}_s^{(n,i)}$$

$$+ \frac{1}{2}\left(\mathbf{H}_s + 2\mathbf{G} + 2\mathbf{E}_d + 4\mathbf{B} + \frac{4}{\gamma\,\Delta t}\mathbf{C}\right)\Delta\boldsymbol{\psi}_d^{(n,i)}$$

$$\tag{9.23b}$$

This equation may now be used to estimate the changes in the potentials and Equation 9.19c to determine the corresponding change in the time derivative term. In each case the required changes in the residuals are the negative of the calculated residuals. The calculation of the term $((2/\gamma\,\Delta t) \times \mathbf{C}R_\psi)^{(n,i)}$ may be incorporated into the calculation $R_d^{(n,i)}$ by replacing $\dot{\psi}_d^{(n,i)}$ with $(\dot{\psi}_d + (1/\gamma\,\Delta t)R_\psi)^{(n,i)}$. As all of the coefficients are independent of the time derivative term this is exactly equivalent to choosing the time derivative term so that Equation 9.18c gives zero residual. Further, since Equation 9.18c is linear, all estimates after the first will always give zero residual so that it is only necessary to ensure that the initial estimate gives zero residual.

The efficiency of a Newton–Raphson solution procedure depends on the choice of the first estimate as well as the method of predicting the required corrections. These corrections should be based on the linear relation with the residuals; thus, the first estimate of $\psi_d^{(n)}$ would be the value from the previous time step since the equations are non-linear in this term.

Once the estimates have been obtained there are three alternatives for calculating the corrections to these estimates. The corrections in Equation 9.23 will change with each iteration and since the calculation of these new terms, and their reduction to triangular factors, is relatively expensive it could be advantageous not to update them at each iteration. The computer program developed for the solution of this problem utilizes a combination of the 'modified N–R' technique, in which the equations are updated once, and the 'full N–R' procedure in which the equations are updated for each iteration. The method is a full N–R procedure until convergence is near and thereafter a modified N–R is used. The decision of when to change from full to modified N–R is automatically made on the following basis. If the number of iterations necessary for convergence with the old equations results in a lower computing time than one iteration with the new equations then the old equations are used. The overall effect of this automatic method is that predominantly full N–R is used when the equations are difficult to solve and modified N–R is used when they are relatively simpler.

In practice the non-symmetric coefficient matrices in Equation 9.23 were excluded, but it is not clear whether the savings in computing time and storage are sufficient to compensate for the decrease in rate of convergence. It is also unnecessary to recalculate the coefficients for each new estimate if the previous coefficients give adequate convergence. The only criterion is that of savings in computing time versus the decreased convergence rate.

9.5 Evaluation of functions

The functions indicated in Figure 9.1 are of primary importance in two-phase immiscible flow and must be represented accurately in the numerical model. This usually involves calculation of not only the function but of one or two derivatives at each integration point every time the residuals or equations are calculated. Allied to this is the requirement that the data be represented smoothly otherwise unnecessary large perturbations occur in residual changes.

Based on these considerations the relative permeability function should be continuous since the function value only enters in the calculation of residuals. Similarly, the saturation versus capillary pressure function should have continuous first derivatives over the whole region since S_0' enters directly in the calculation of residuals. It could also be argued that convergence would improve by increasing the continuity requirement by one additional derivative since the coefficients of the tangential equations contain one higher derivative. However, this is not considered nearly as important as requiring continuity of the residuals.

In the case of the relative permeability functions a good representation is achieved by linear interpolation between equally spaced points over the full range of saturation. A better representation of the curves would be obtained simply by increasing the number of points without unduly affecting the total computer time. If a large number of points are used care must be taken to ensure that the first derivative, or first differences, are accurately represented.

The saturation/capillary pressure curve is represented by curve fitting a parabola through each set of three adjacent points and using a weighted average of the two parabolas between each two points. The weighting for each parabola varies between unity at the mid-point to zero at both end points. This is equivalent to cubic Hermitian interpolation between each pair of points with the specified slopes at each point taken as the average of the chord slopes at that point. The end slopes may be specified independently and again the use of equally spaced points greatly reduces the computational difficulty.

The range of capillary pressure for the saturation curve may be arbitrary and outside this range linear extrapolation is used, based on the specified slopes at the end point. It is usual to specify non-zero positive end slopes so that the contribution to the matrix **C** for any element does not become zero and result in singular equations.

9.6 Numerical examples

Based on the general approach outlined, a computer program was written to test the method with linear, parabolic or cubic quadrilateral isoparametric element. The program was first used on a one-dimensional problem for which the isoparametric elements became simple linear, parabolic, or cubic interpolation polynomials. A typical isoparametric element is shown in Figure 9.2 and the reader is referred to the text by Zienkiewicz[4] for details of such elements.

9 6.1 One-dimensional problem

The problem chosen is one that had previously been solved with a Galerkin approach using cubic splines[1] and with the moving reference point method.[2] The length of the region is 45·7 cm, the porosity of the rock is 0·46 and the permeability is $0·355 \times 10^{-6}$ cm^2. The oil viscosity is 1·65 poise and the water viscosity 0·38 poise. The relative permeability and capillary pressure

Finite Elements in Fluids

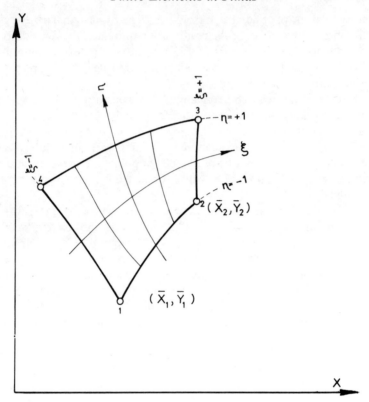

Figure 9.2 Typical isoparametric element

curves are plotted against saturation in Figure 9.1. The residual oil value is 10 per cent and the residual water value 15 per cent.

The initial conditions are those of zero flow and 15 per cent (residual) water saturation, the rest of the pore space being saturated by the oil. The boundary conditions at one end are a specified injection rate of water and zero flow of oil. At the other end, the sum of the pressures is specified as being zero and the ratio of the flows equals the ratio of the mobilities. This is equivalent to saying that the slope of the saturation curve is zero at that end.

The effect of capillary forces alone in a homogeneous porous medium will be to create a flow which will eventually result in a uniform distribution of oil and water. If water is injected at one end a saturation 'front' develops which is smooth for low flow rates but gets progressively sharper as flow rate increases. Two flow rates were analysed which gave extreme conditions from smooth to sharp fronts as quoted in Reference 1.

The elements were chosen of equal length, with equally spaced interior nodes, and the integration over each element was carried out by a three-point Gaussian quadrature rule. In using the previously outlined method for integration with respect to time, a constant value of γ is used throughout. A specified time domain t is automatically split into sub-intervals (time steps) of the form $\Delta t/2^n$ where n may be different for each time step. If $\dot{S}_{max}$ is the maximum rate of change of saturation at the end of a time step, the factor n is chosen so that $(\Delta t/2^n)\dot{S}_{max} \leqslant \Delta S$ for that time step. When $(\Delta t/2^n)\dot{S}_{max} \leqslant 0.4 \Delta S$ then the integer n is decreased by one for the next time step.

The low flow rate of 0.776×10^{-4} cm/s produced the smooth saturation distribution shown in Figure 9.3 and the results can be seen to agree well with results obtained by the Galerkin approach with cubic splines.[1]

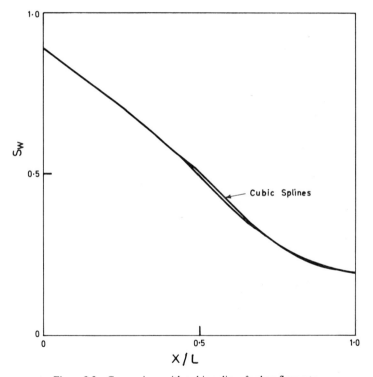

Figure 9.3 Comparison with cubic splines for low flow rate

The moving reference point method and the centred-in-distance finite-difference method with 20 intervals[2] also give good results for this relatively easy problem. The calculation was carried out with seven parabolic elements giving a total of 15 degrees of freedom. Integration with respect to time was

Finite Elements in Fluids

executed for a value of $\gamma = 0.5$, a time interval corresponding to 0·01 pore-volumes and a maximum saturation change of $S = 0.05$. At a total of 0·36 pore-volumes the actual number of time steps taken was 59 giving an average time step corresponding to 0·0061 pore-volumes.

The high flow rate of 0·776 × 10⁻² cm/s was more difficult to solve because of the very steep saturation front as shown by the solutions obtained in References 1 and 2. The injection was started with 0·025 pore-volumes of water at the low flow rate and then increased to the high flow rate to a total injection of 0·28 pore-volumes in order to compare with previous results.[1] Integration with respect to time was carried out without restricting the time step but specifying a maximum saturation change ΔS. Thus, the first time interval corresponded to 0·025 pore-volumes but commenced with a sub-interval of 1/32nd. The second time interval corresponded to the remaining 0·255 pore volumes and started with a sub-interval of 1/128th.

The problem was run keeping a constant $\gamma = 0.5$ and values of ΔS of 0·05, 0·10 and 0·20 with little variation in the results. For $\Delta S = 0.20$, the actual average value of S was approximately 0·084 for a total of 31 time steps required to reach 0·28 pore-volumes. As the numerical integration depends

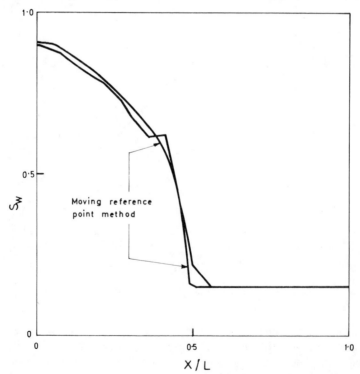

Figure 9.4 Comparison with moving reference point method for high flow rate

only on the values at the Gauss points, the results are plotted as straight lines connecting these values. A comparison of these results and those obtained by the moving reference point method[1] is shown in Figure 9.4 and a comparison with the cubic-spline solution[1] for the case of twelve increments (15 degrees of freedom) is shown in Figure 9.5. While the three methods show good agreement, the finite difference method[2] with 21 degrees of freedom gave considerably less accurate results at 0·36 pore-volumes injected for the same problem (also supposedly at less computational effort).

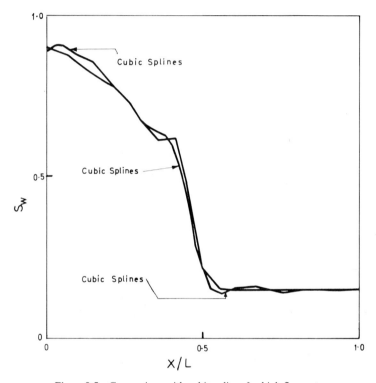

Figure 9.5   Comparison with cubic splines for high flow rate

9.6.2   Two-dimensional problem

A hypothetical two-dimensional problem was chosen with a well-bore radius of 0·08 metres and an internal radius of drainage of 324 metres. The details of the element subdivision are given in Figure 9.6 along with the location of the producing zones and the original oil/water contact.

The total thickness of the reservoir was 50·0 metres and for the computer calculation the model was sectioned into seventy five elements, five in the

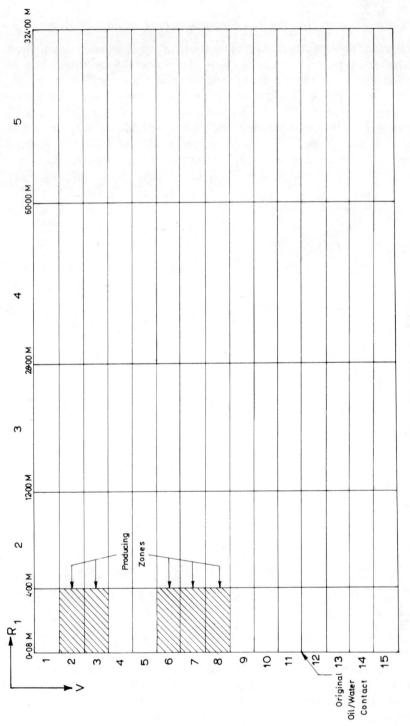

Figure 9.6 Finite element mesh

radial and fifteen in the vertical direction as shown in Figure 9.6. The influx of water was assumed to occur uniformly across the base of the reservoir with no-flow occurring at the radius of drainage boundary. The oil was assumed to be produced from two perforated zones represented by elements two, three, six, seven and eight. The thickness of these zones were 4·2 and 3·2 metres respectively.

The relative permeability and capillary pressure curves for this hypothetical reservoir are similar to those in Figure 9.1. It is realized that the solution obtained with these particular functions is not unique and that many varying combinations of saturation functions could well achieve similar results.

The ratio of oil and water flow at the wellbore is a function of the ratio of the mobilities at these perforated zones. A condition of zero flow is specified along the remaining boundaries except for the reservoir/aquifer contact at which an influx of water occurs.

During the course of production the invading water reacts with the oil zone thus changing the geometry of the original oil/water contact. It is well known that if the injection of water exceeds a certain flow rate then 'coning' of the well occurs. This phenomenon of coning manifests itself in the form of a conical mass of water rising rapidly towards the producing zones. Once the water enters the producing zones the efficiency of the well decreases as water is now produced with the oil and has to be separated at the surface. It eventually becomes uneconomical to produce the well as the ratio of water to oil production becomes significant. This deleterious effect is highly undesirable, especially since there is very little hysteresis and so large quantities of oil could be trapped in the reservoir. This analysis was concerned with matching the original reservoir performance to establish the viability of the finite element model. Linear isoparametric elements were used with a two-point Gaussian quadrature rule in both vertical and radial directions. After the initial time step an estimate is made of the saturation change to be expected in the next time interval. If this exceeds a set value then the time step is reduced to comply with this tolerance. Alternatively the time step could be increased if the tolerance was greatly underestimated. The objective is to proceed with the time-stepping procedure as economically as possible.

The results achieved from the analysis are presented in Figure 9.7. The production of oil and water as predicted by the finite element analysis is compared with the hypothetical field results and are seen to be very favourable. As stated previously the matching of the curves is certainly not unique and could have been achieved by varying other of the material properties. However, the aim of the study was to construct a working two-dimensional two-phase model and this goal was satisfactorily reached.

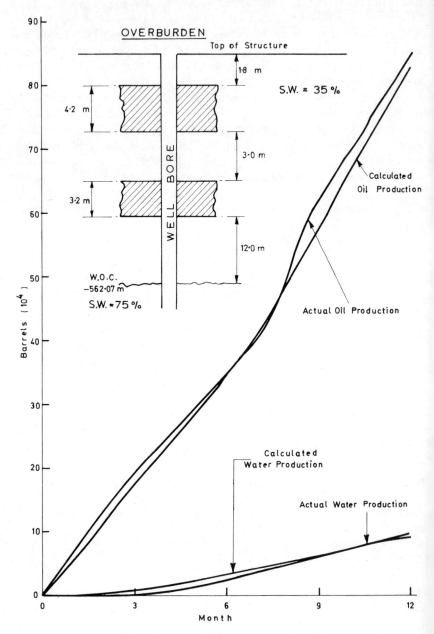

Figure 9.7   Two-dimensional history match

9.7 Conclusion

The numerical computations presented here show that the piecewise parabolic finite element spatial approximation is competitive with the cubic splines method. In fact it appears that the high continuity of the cubic splines may cause more of a precursor in front of the saturation front than is obtained with the parabolic elements. However, it appears that the parabolic elements will exhibit a less smooth solution behind the front. This could be caused by the occurrence of very sharp cusps between two adjacent parabolas but was found to be of little consequence in the present problem. Although these cusps also occur with linear elements they are less acute and would probably give better behaviour in this respect.

The two-dimensional coning problem was solved using linear finite elements and the results were encouraging in that a reasonable history match was obtained. It is difficult to draw comparisons between the two forms of solution used in this problem as entirely different computers were used. However, the results indicate that the finite element method is certaintly a viable technique in the solution of such problems.

Acknowledgements

The authors wish to thank the British Petroleum Company for their financial support to this research project.

References

1. J. Douglas, Jr., T. Dupont and H. H. Rachford, 'The application of variational methods to waterflooding problems' *The Journal of Canadian Petroleum Technology*, pp. 79–85 (September, 1969).
2. H. H. Rachford, 'Numerical calculation of immiscible displacement by a moving reference point method', *Society of Petroleum Engineers Journal*, 87–101 (June, 1966).
3. W. Liniger and R. A. Willoughby, 'Efficient methods for still systems of ordinary differential equations', *SIAM Journal of Numerical Analysis*', 7, 1, 47–66 (1970).
4. O. C. Zienkiewicz, *The Finite Element Method in Engineering Science*, McGraw-Hill, 1971.

Chapter 10

Galerkin Approach to Saturated–Unsaturated Flow in Porous Media

S. P. Neuman

10.1 Introduction

Problems of seepage in unsaturated and partly saturated porous media lead to quasilinear partial differential equations that are extremely difficult to solve by analytical methods. Many attempts to solve these equations by finite difference techniques have been reported in the literature, primarily by soil scientists. A comprehensive review of the finite difference approach as well as a complete survey of all pertinent literature were given by the author in a recent report.[2]

In dealing with non-uniform systems of complex geometry and arbitrary anisotropy, the finite difference approach is sometimes difficult to apply. The treatment of inclined atmospheric boundaries such as seepage faces and evaporation surfaces is difficult with this approach because of the way that prescribed flux boundaries are handled. Such boundaries usually lead to a non-symmetric finite difference matrix, which is a disadvantage if one wants to solve the equations simultaneously at all nodes with the aid of a direct method such as the Gauss elimination scheme. Verma and Brutsaert[13] found that the highly non-linear saturated–unsaturated flow problem often converges faster if Gauss elimination is employed over all nodes simultaneously than if other, less implicit iterative procedures are used.

Perhaps one of the most important features of the finite element approach is the ability of the grid to contract or expand at any stage of the computation to follow the contours of moving boundaries, free surfaces and deformable material interfaces.[4,10,11] This may be important when unsaturated or partly saturated conditions are encountered as, for example in compressible peat deposits, swelling clays or fractured rock formations. The technique described in the present study can be readily extended to account for such situations.

In the present work the problem of saturated–unsaturated seepage is solved by a Galerkin-type element method in conjunction with a fully implicit iterative scheme. The resulting computer program can handle plane and axisymmetric flow regions having complex boundaries, arbitrary degrees of non-uniformity and anisotropy of any kind. Boundary conditions along seepage faces and evaporation surfaces are easily simulated in a manner that preserves the symmetric nature of the finite element matrix. Flow to a fully or partially penetrating well is handled by taking into account well storage as well as the characteristics of the pump. Experience with the finite element algorithm indicates that convergence of the iterative scheme is extremely fast in many cases. Examples are included which show that the classical concept of a 'free surface' is unsuitable for many engineering problems.

10.2   Description of the problem

The flow of water in an unsaturated or partly saturated porous medium can be described by (see Reference 9)

$$\mathscr{D}(\psi) = \nabla \cdot \{KK_r\nabla(\psi + z)\} - \left(C_s + \frac{\theta}{v}S_s\right)\frac{\partial\psi}{\partial t} = 0 \qquad (10.1)$$

where $\mathscr{D}$ is a quasilinear differential operator defined in the flow region, ψ is pressure head, z is elevation head, t is time, θ is volumetric moisture content (fraction of bulk volume), v is porosity, $\mathbf{K}$ is hydraulic conductivity tensor at saturation, K_r is relative hydraulic conductivity ($0 < K_r \leqslant 1$), C_s is specific moisture capacity (defined as $\partial\theta/\partial\psi$), v is porosity and S_s is specific storage (defined as the volume of water instantaneously released from storage per unit bulk volume when ψ is lowered by one unit).

The quantities $\mathbf{K}$, v and S_s are functions of position only. The term S_s reflects the combined elastic properties of the medium and the water when one is willing to assume that lateral strains are negligible and that the total stress at each point remains fixed in time. The exact definition of S_s for fully saturated media can be found in the recent work of Gambolati[5] who shows[6] that its use as a constant is justified as long as vertical strains do not exceed 5 per cent. The applicability of this concept to unsaturated conditions requires further investigation. In the present work it is assumed that S_s can be disregarded in the unsaturated zone because the effect of compressibility on the storage of water is negligibly small in comparison to the effect of changes in the moisture content, θ.

The pressure head, ψ, is taken to be positive in the saturated zone and negative in the unsaturated zone. In the absence of hysteresis effects, K_r and ψ are monotonically increasing single-valued functions of θ, and C is the derivative of θ with respect to ψ. The functional relationships between K_r, ψ, C and θ are different for each soil and must usually be determined by field[7] or laboratory[1] experiments. The nature of these relationships is often highly non-linear.

Equation 10.1 must be supplemented by the initial conditions

$$\psi(x_i, 0) = \psi_0(x_i) \tag{10.2}$$

at each point in the interior of the flow region represented by the Cartesian coordinates x_i ($i = 1, 2, 3$). In addition, one must specify either the pressure head or the normal flux at each point along the boundary. If Γ_1 is the segment of the boundary, Γ, along which pressure heads are prescribed, and Γ_2 is the segment along which normal fluxes are prescribed, then the boundary conditions become

$$\psi(x_i, t) = \Psi(x_i, t) \quad \text{on } \Gamma_1 \tag{10.3}$$

$$\{\mathbf{K}K_r\nabla(\psi + z)\} \cdot \mathbf{n} = -V(x_i, t) \quad \text{on } \Gamma_2 \tag{10.4}$$

Here Ψ and V are known functions of x_i and t, and $\mathbf{n}$ is the unit outer normal on Γ.

10.3 Numerical approach

The present work adopts a network of triangular elements for plane flow and of concentric rings of constant triangular cross-section for axisymmetric problems. In an individual element the approximation of the pressure head, $\bar{\psi}$, is described in terms of shape functions N_i and the values ψ_i of the pressure head at the element corner (node) points

$$\bar{\psi} = \sum_{i=1}^{n^e} \mathbf{N}_i^e \psi_i = \mathbf{N}^e \boldsymbol{\psi} \tag{10.5}$$

where n^e denotes the number of element node points. In accordance with the Galerkin method these values ψ_i must be determined so as to satisfy the initial and boundary conditions of the problem together with the orthogonality requirement

$$\sum_e \int_{\Omega^e} \mathscr{D}(\psi^n) N_i^e \, d\Omega = 0; \qquad i = 1, 2, \ldots, n \tag{10.6}$$

where Ω^e is the interior of element e and n is the total number of nodes in the flow region.

Since the Galerkin method applies only at a given instant of time, the time derivative $\partial\psi/\partial t = \dot{\psi}$ appearing in $\mathscr{D}(\psi)$ must be determined independently of the orthogonalization process. It was found that if $\dot{\psi}$ is replaced by $\dot{\bar{\psi}}$ as is often done in the finite element approach, the calculated pressure head, $\bar{\psi}$, tends to exhibit spatial oscillations around its limit. Owing to the nature of the non-linear relationships between ψ, θ, K_r and C_s, these oscillations may often prevent the solution from converging in the case of unsaturated flow. The rate of convergence can be improved dramatically by defining the nodal values of the time derivatives as weighted averages over the entire flow region according to

$$\frac{\partial \psi_i}{\partial t} = \frac{\sum_e \int_{\Omega^e} (C_s + (\theta/v)S_s) \, \partial \psi/\partial t \, N_i^e \, d\Omega}{\sum_e \int_{\Omega^e} (C_s + (\theta/v)S_s) N_i^e \, d\Omega} \tag{10.7}$$

In addition to speeding up the rate of convergence, Equation 10.7 also results in a considerable increase in computational efficiency during each iteration, as will be demonstrated below.

Let us assume that $\mathbf{K}$, v and S_s are constant in each element while K_r, C and θ vary linearly between the nodes. Then, by applying Green's first identity to Equation 10.6 one obtains a set of quasilinear first-order differential equations,

$$\mathbf{A}\psi + \mathbf{C}\dot{\psi} = \mathbf{Q} - \mathbf{B} \tag{10.8}$$

Here $\mathbf{A}$ is a sparse symmetric $n \times n$ matrix depending only on geometry and hydraulic conductivity,

$$\mathbf{A} = \sum_e \int_{\Omega^e} \mathbf{K}_r^{\mathrm{T}} \mathbf{N}^e \nabla \mathbf{N}^{e\mathrm{T}} \mathbf{K} \nabla \mathbf{N}^e \, d\Omega \tag{10.9}$$

where $\mathbf{K}_r$ and $\mathbf{N}^e$ are n-dimensional vectors with three non-zero components represented by the nodal values of K_r and N^e in e, respectively; $\nabla \mathbf{N}^e$ is an $n \times 2$ matrix with 3×2 non-zero components representing the gradient of $\mathbf{N}^e$ with respect to the two global space coordinates in the plane of the triangle. $\mathbf{C}$ is a diagonal $n \times n$ matrix depending on geometry as well as on C_m, θ, S_s and v,

$$\mathbf{C} = \sum_e \int_{\Omega^e} \left(\mathbf{C}_s + \frac{S_s}{v}\theta \right) \mathbf{N}^{e\mathrm{T}} \mathbf{N}^e \, d\Omega \tag{10.10}$$

where C_m and θ are n-dimensional vectors just like $\mathbf{K}_r$ and $\mathbf{N}^e$; $\mathbf{Q}$ is an n-dimensional vector representing nodal fluxes into or out of the system,

$$\mathbf{Q} = -\sum_e \int_{\Gamma^e} \nabla \mathbf{N}^e \, d\Gamma \tag{10.11}$$

which vanishes at nodes that do not act as sources or sinks; and $\mathbf{B}$ is an n-dimensional vector depending solely on geometry and hydraulic conductivity,

$$\mathbf{B} = \sum_e \int_{\Omega^e} \mathbf{K}_r^{\mathrm{T}} \mathbf{N}^e K_z \nabla \mathbf{N}^e \, d\Omega \tag{10.12}$$

where $\mathbf{K}_z$ represents only those components of $\mathbf{K}$ which include the vertical direction, implying that $\mathbf{B}$ must vanish in the case of horizontal plane flow.

The diagonal nature of **C** is a consequence of the averaging process performed in Equation 10.7. It enables one to further reduce computational effort and storage requirements by combining three or four triangular elements into a single triangle or quadrilateral and by eliminating the equation for the midpoint. This is accomplished merely by first redistributing the values of **C** and **Q** between the peripheral nodes, in analogy to their relative distribution in the original triangle, and then assigning zero values to these terms at the midpoint. The equation for the midpoint can then be solved explicitly for ψ_{midpoint} and the result substituted in the equations for the peripheral nodes. In this manner the total number of nodal points is greatly reduced without impairing the symmetric nature of **A**, while the accuracy of the interpolation scheme remains essentially unchanged.

10.4   Iterative procedure

In the particular case where the porous medium remains unsaturated at all times, Equation 10.8 can be replaced by the Crank–Nicolson finite difference scheme

$$\left(A + \frac{2}{\Delta t}C\right)^{k+\frac{1}{2}}\psi^{k+1} = 2(Q - B)^{k+\frac{1}{2}} - \left(A - \frac{2}{\Delta t}C\right)^{k+\frac{1}{2}}\psi^{k} \tag{10.13}$$

where k represents time steps and $\Delta t = t^{k+1} - t^{k}$. The coefficients in Equation 10.13 are evaluated at half the time step on the basis of $\psi^{k+\frac{1}{2}}$. A first estimate of $\psi^{k+\frac{1}{2}}$ can be obtained either from an explicit formulation of Equation 10.8,

$$\psi^{k+\frac{1}{2}} = \left(1 - \frac{\Delta t}{2}C^{-1}A\right)^{k}\psi^{k} + \frac{\Delta t}{2}(C^{-1}Q - C^{-1}B)^{k} \tag{10.14}$$

or by linear extrapolation from ψ^{k-1} and ψ^{k}. Experience has shown that the explicit predictor has no apparent advantages in terms of convergence rate over simple linear extrapolation and therefore the latter approach has been adopted for this work.

Equation 10.13 is solved by Gauss elimination for ψ^{k+1} and the results are used to obtain an improved estimate of the coefficients on the basis of $\psi^{k+\frac{1}{2}} = (\psi^{k} + \psi^{k+1})/2$. Since this iterative procedure is fully implicit, it usually requires only a few iterations to converge. Experiments with a fully implicit quasi-linearization (Newton–Raphson) scheme have failed to produce comparable results. This may probably be due to the high sensitivity of the Newton-Raphson method to the initial estimate of $\psi^{k+\frac{1}{2}}$ and to the increased amount of computational effort required for each iteration.

If part of the porous medium is saturated and S_{s} is zero, **C** in the saturated zone vanishes and the governing equation there becomes elliptic. This means that sudden changes in any of the boundary conditions around the saturated zone have an instantaneous effect on ψ everywhere inside this zone, so that ψ ceases to be a continuous function of time. Consequently, the right-hand side of Equation 10.13 becomes unknown and the problem cannot be solved. This difficulty can be overcome by adopting the backward difference scheme

$$\left(\mathbf{A} + \frac{1}{\Delta t}\mathbf{C}\right)^{k+\frac{1}{2}}\boldsymbol{\psi}^{k+1} = (\mathbf{Q} - \mathbf{B})^{k+\frac{1}{2}} + \frac{1}{\Delta t}\mathbf{C}^{k+\frac{1}{2}}\boldsymbol{\psi}^{k} \qquad (10.15)$$

where the last term on the right-hand side vanishes in the saturated zone. The effect of evaluating the coefficients in Equation 10.15 at half the time step is to under-relax the system. Experience with highly non-linear systems of equations such as Equation 10.8 indicates that under-relaxation may often be necessary to obtain a satisfactory rate of convergence (see for example the work of Cooley[3]), and this has been confirmed by numerical experiments conducted during the present study.

10.5 Boundary conditions at seepage faces

A seepage face, as the term implies, is an external boundary of the saturated zone where water seeps out of the porous medium and therefore the pressure head is uniformly zero (assuming that atmospheric pressure is zero). One of the major difficulties in dealing with partly-saturated flow problems stems from the fact that the length of the seepage face varies during each time step in a manner that cannot be predicted *a priori*. This difficulty has been overcome in the present work owing to the ease with which prescribed pressure head and prescribed normal flux boundary conditions can be assigned at each node with the finite element method. The proposed iterative procedure would be quite cumbersome to use with conventional finite difference techniques, particularly in anisotropic media with irregularly shaped seepage faces, because prescribed flux boundaries are relatively difficult to handle.

Let us consider a given segment of the boundary along which a seepage face has chances to develop during any stage of the computation. During each iteration, the saturated part of this segment is treated as a prescribed pressure head boundary with $\psi = 0$. At the same time, the unsaturated part is treated as a prescribed flux boundary with $Q = 0$. The relative length of each part is continually adjusted during the iterative process until all calculated values of Q along the saturated part and all calculated values of ψ along the unsaturated part are negative, indicating that water is leaving the porous medium through the saturated part of the boundary.

10.6 Flow to a well

When a well has been completed in an unconfined aquifer and discharges at a rate $Q_D(t)$, flow into the wellbore is not uniform along its length. As shown in Figure 10.1, the boundaries along the wellbore consist of an upper segment (Γ_2) across which no water can flow into the well due to the unsaturated state of the porous medium, a seepage face S, and a boundary Γ_1 where the total head at any instant of time is uniform and equal to the elevation of the

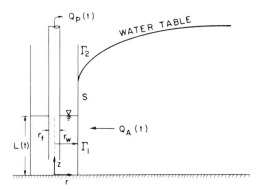

Figure 10.1

water level, $L(t)$. The total discharge from the pump, $Q_P(t)$, consists of two components, the discharge from the aquifer into the well, $Q_A(t)$, and the amount of discharge contributed from well storage.

If one assumes that L, Q_P and Q_A vary linearly during each time step, then a material balance calculation for the well leads to

$$\Delta L = \frac{\Delta t}{\pi(r_w^2 - r_t^2)}(Q_P^{k+\frac{1}{2}} - Q_A^{k+\frac{1}{2}}) \qquad (10.16)$$

Here ΔL is the change in the height of the water level in the well during Δt, r_w is the effective radius of the well, and r_t is the outside radius of the production pipe.

The total discharge from the pump can be regulated at the ground surface by controlling the capacity of the pump. In this case, $Q_P(t)$ is a prescribed function of time and Equation 10.16 involves only two unknowns, ΔL and $Q_A^{k+\frac{1}{2}}$. Another possibility is to set the pump operation at a given maximum capacity, Q_M, and allow $Q_P(t)$ to vary with $L(t)$ according to

$$Q_P(t) = f(L) \leqslant Q_M \qquad (10.17)$$

where $f(L)$ is a prescribed function of L. In a given pump–aquifer system, the shape of $f(L)$ can be determined from the characteristic discharge curves of the pump and the design specifications of the installation. When Equation 10.17 is inserted into Equation 10.16 one obtains

$$\Delta L = \frac{\Delta t}{\pi(r_w^2 - r_t^2)}\left[f\left(L^k + \frac{\Delta L}{2}\right) - Q_A^{k+\frac{1}{2}} \right] \qquad (10.18)$$

which again involves the two unknowns ΔL and $Q_A^{k+\frac{1}{2}}$.

During each time step, an estimate of ΔL for the first iteration is obtained from Equation 10.16 or Equation 10.18 on the basis of $Q_A^{k+\frac{1}{2}}$ and $f(L^{k-\frac{1}{2}})$, and $L_{old}^{k+\frac{1}{2}}$ in Equation 10.19 below is set equal to L^k. The average value of the water level in the well for the time step is then adjusted according to

$$L_{new}^{k+\frac{1}{2}} = \alpha L_{old}^{k+\frac{1}{2}} + (1 + \alpha)(L^k + \Delta L); \qquad 0\cdot5 \leqslant \alpha \leqslant 1 \qquad (10.19)$$

where α is an under-relaxation factor. Experience has shown that α should increase as Q_A approaches Q_P and should usually exceed $0\cdot7$. During subsequent iterations, Equations 10.16 or 10.18 are used repeatedly with Equation 10.19 until a desired degree of convergence is achieved for ψ at all nodes. At the end of the time step, the value of L^{k+1} is calculated as $L_{new}^{k+\frac{1}{2}} + \Delta L/2$.

10.7 Evaporation and infiltration boundary conditions

Bare soils can lose water to the atmosphere by evaporation or gain water by infiltration due to rainfall or sprinkler irrigation. While the potential (i.e. maximum possible) rate of evaporation from a given soil depends only on atmospheric conditions, the actual flux across the soil surface is limited by the ability of the porous medium to transmit water from below. Similarly, if the rain intensity (i.e. potential rate of infiltration) exceeds the infiltration capacity of the soil, part of the rain may be lost by run-off. Here, again, the potential rate of infiltration is controlled by atmospheric conditions, while the actual flux is limited by the infiltration capacity, which depends on antecedent moisture conditions in the soil.

Thus, the exact boundary conditions at the soil surface cannot be predicted *a priori* and a solution must be sought by maximizing the absolute value of the flux subject to the requirements

$$|\{KK_r\nabla(\psi + z)\} \cdot \mathbf{n}| \leqslant |V_P| \qquad (10.20)$$

$$\psi_L \leqslant \psi \leqslant 0 \qquad (10.21)$$

where V_P is the prescribed potential flux and ψ_L is the minimum allowed pressure head at the soil surface. Equation 10.21 states that, in the absence of water accumulation at the soil surface, ψ is limited by ψ_L from below and by atmospheric pressure from above. The lower limit, ψ_L, can be determined from equilibrium conditions between soil water and atmospheric vapour (see, for example, Reference 12, p. 157).

The finite element method is exceptionally well suited for the treatment of Equations 10.20 and 10.21 because the type of boundary condition can be easily changed at each node from one iteration to another. During the first iteration in each time step, the surface nodes are treated as a prescribed flux boundary and are assigned an arbitrary fraction of the potential flux, usually 0·1. If the computed values of ψ_n satisfy Equation 10.21, the absolute value of the flux at each node, n, is increased by $|\psi_L|/|\psi_n|$ in the case of evaporation, or by $|\psi_L|/|\psi_L - \psi_n|$ in the case of infiltration, subject to Equation 10.20. If some value of ψ_n lies outside the limits specified by Equation 10.21, then, during the subsequent iteration, n is treated as a prescribed pressure head node with $\psi = \psi_L$ for evaporation or $\psi = 0$ for infiltration. This situation is maintained as long as Equation 10.20 is satisfied. If, at any stage of the computation, the calculated flux exceeds the potential flux so that Equation 10.20 is not satisfied, n is assigned the potential flux and is again treated as a prescribed flux boundary. The iterative procedure continues until convergence is achieved at all nodes in the finite element network.

10.8 Examples

(1) As a first example, consider an earth dam with a sloping core and a horizontal drainage blanket. A cross-section of the dam, together with the superimposed finite element network, are shown in Figure 10.2. The saturated hydraulic conductivity in the sandy shell material is 0·5 cm per unit time in the horizontal direction and 0·1 cm per unit time in the vertical direction (arbitrary time units are used). In the clay core, the saturated conductivity is 0·0001 cm per unit time in the horizontal direction and 0·001 cm per unit time in the vertical direction. The unsaturated properties of both materials are shown in Figure 10.3. Evaporation at the surface is suppressed and S_s is equal to zero.

As earth dam fills are sometimes compacted at relatively high moisture contents, the initial values of θ are taken to be 0·255 in the shell and 0·598 in the core. At time $t = 0$, the water level in the reservoir is suddenly raised to 4 metres. At $t = 182$, the water level starts rising again at a rate of 1 metre per 24 time units, reaching an elevation of 12 metres at $t = 374$. This situation is maintained indefinitely.

Figure 10.4 shows how the so called 'free surface', corresponding·to $\psi = 0$, advances with time. In the classical approach this surface is treated as a moving material boundary, whereas in the present work it is merely an internal isobar which happens to separate the saturated and unsaturated portions of the dam. It is seen that this surface may have an inverted shape, a situation which cannot be handled with the classical 'free surface' approach. In addition, the rate of advance of the zero-pressure surface cannot be correctly predicted by a method which ignores antecedent moisture conditions in the unsaturated zone.

Accumulation of water on the upstream face of the clay core and the formation of a saturated mound at the bottom of the downstream shell are caused by drainage from the unsaturated zone. The rate of this drainage

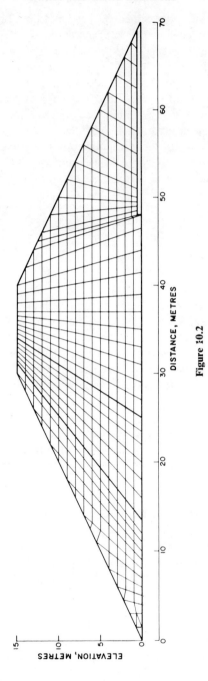

Figure 10.2

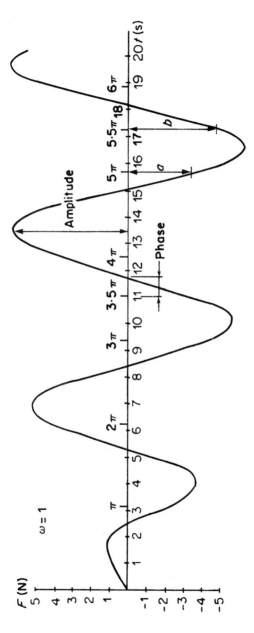

Figure 12.4 Hydrodynamic force on the body

has been used to guarantee continuity up to and including the first derivative. This prescribed vertical velocity is given by

$$u_z = 0 \qquad\qquad \text{for } t < 0$$

$$u_z = \frac{\omega^3}{\pi^2}\left(t^3 - \frac{\pi}{\omega}t^2\right) \quad \text{for } 0 < t < \pi/\omega \qquad (12.27)$$

$$u_z = -\sin(\omega t) \qquad \text{for } t > \pi/\omega$$

An example of the response in terms of the hydrodynamic force due to the hydrodynamic pressure $-\rho\,\partial\phi/\partial t$ is given in Figure 12.4. The calculated force on the body is approximately equal to

$$F = \hat{M}\omega\cos(\omega t) + \hat{C}\sin(\omega t) \qquad (12.28)$$

where $\hat{M}$ is the added mass coefficient and $\hat{C}$ the damping coefficient. The values of $\hat{M}\omega$ and $\hat{C}$ are found for those values of t where $\sin(\omega t)$ and $\cos(\omega t)$ are respectively zero. However, $\hat{M}\omega$ and $\hat{C}$ can also be calculated by means of the amplitude and the phase of the response. In this determination of $\hat{M}$ and $\hat{C}$ the first oscillation is to be avoided, because of initialization of the response, whereas after a number of oscillations the reflection of the waves will cause the end of a meaningful response.

The results for $\hat{M}$ and $\hat{C}$ have been compared with those of Vught[11] who measured and calculated these coefficients for deep water. These data are collected in Table 12.1. For $\omega = 1$ the basin is to be considered shallow relative to the wave length, which explains the significant differences in the results of the first column.

Table 12.1   Hydrodynamic coefficients

	Frequency ω (rad/s)	1	2	3	4
Added mass $\hat{M}$ in kg/m	finite method	4000	1600	1500	1600
	Reference 11	2800	1600	1500	1600
Added damping $\hat{C}$ in kg/m . s	finite method	4100	3100	1900	900
	Reference 11	2800	3100	1900	900

12.4.2 Wave length, wave and group velocity

The model drawn in Figure 12.3 has also been used for checking some wave properties. The motion of the body is described by Equation 12.27 but after three oscillations the body is kept at rest. The numerical experiment was carried out for $\omega = 1\cdot5$ and a time range of 30 seconds real time. From observation of the wave height distribution at various time intervals it has been concluded that the results for wave length, wave velocity and group velocity agreed within 3 per cent with the analytical solutions.

12.4.3 Curvature of the floating body

The curvature of a two-dimensional floating body is represented by straight lines. The original rectangular model in Figure 12.3 has been replaced by the 'curved' model of Figure 12.5. The element distributions in the fluid has also been changed near the hull of the body.

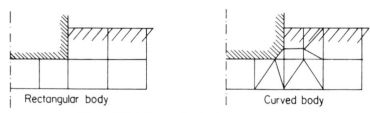

Rectangular body	Curved body

Figure 12.5 Element distribution near the body

The hydrodynamic coefficients derived for the two bodies differed by no more than 2–4 per cent. However, the time-step in the Runge–Kutta integration procedure was much smaller for the curved model with the small elements than for the original model. This result confirmed the estimate for the time-step given in Equation 12.24.

12.4.4 Wave tank experiments

In many cases the problems accessible with the finite element method resemble the modelling as used in wave tank experiments. To a large extent this resemblance is caused by the finite domain of the finite element model.

The example described in Section 12.4.1 deals with the determination of hydrodynamic coefficients of a floating body. These coefficients could well be determined when the experiment lasted at least four oscillations without possible interference of reflected waves. Combination of the time necessary for four oscillations with the groups velocity (12.26) would lead to a distance between floating object and boundary of the wave tank of at least one wave length. However, it has been found that a minimum distance of 1·5 times the wave length is much more appropriate.

The wave generator is another component of a wave tank. The prescribed normal velocity at the wave generator can be incorporated in the finite element model. For experiments of a single frequency the velocity distribution at the wave generator can be taken from given analytical solutions of the velocity distribution in deep or shallow water. However, a linear distribution of the prescribed velocity at the wave generator will lead to equally accurate results and such a distribution can also be applied when more complicated wave distributions are to be generated.

In the finite element model it is impossible to construct a wall without reflection. In a wave tank the reflection is reduced by a beach at the end of the tank. The representation of the slope of a beach in a finite element model is no problem. The damping of an actual beach can also be simulated by introducing surface damping over part of the free surface. Surface damping per unit area proportional to the time derivative of the wave height changes the dynamic boundary condition as follows

$$\frac{\partial \phi}{\partial t} + g\eta + c\frac{\partial \eta}{\partial t} = 0 \qquad (12.29)$$

where c is a damping coefficient. In the functional Π this damping is represented by an extra term on a part Γ_{nd} of the free surface S_η

$$\int_{\Gamma_{nd}} c \cdot \eta \frac{\partial \eta}{\partial t} \, d\Gamma \qquad (12.30)$$

This extra term has no effect on the numbers of unknowns in Equation 12.19. In the problem shown in Figure 12.3, for $\omega = 1.5$, a surface damping of 1 m/s over 20 m of the surface adequately dissipates the wave energy. Higher values of c lead to additional reflection due to the damping surface itself.

Studies of irregular motions of a floating body in a wave tank are usually of long duration. However, the tools explained in this section allow a successful, numerical simulation of such experiments.

12.4.5 Three-dimensional problems

A first extension of the application of the method lies in the solution of motion problems of axially symmetric, SPAR-type floating structures. The application to truly three-dimensional problems is restricted by the capabilities of the computer.

To illustrate this point, let us consider a three-dimensional, box-type floating body of two axes of symmetry. For the determination of the hydrodynamic coefficients, similar to the example drawn in Figure 12.3 and described in Section 12.4.1, the fluid domain should satisfy the following criteria. From Sections 12.2.2 and 12.4.4 it can be concluded that the (deep) water domain should be at least $0.5l$ deep and $1.5l$ wide, where the wavelength l is related to the frequency ω for which the calculation is to be carried out by $l = 2\pi g/\omega^2$. Application of the estimate for the maximum element length derived in Section 12.3.1 leads to a division of the fluid domain into at least 8×8 elements in the horizontal plane and into 2 to 3 elements in the vertical direction.

The computer time necessary to eliminate potentials in the interior points of the fluid domain according to Section 12.3.2 and to solve the set of differential equations with the Runge–Kutta method is approximately 15 and 5 minutes CPU on an IBM 360/75 computer, respectively.

The simulation of a similar wave tank experiment to study the irregular motions of a three-dimensional body needs a fluid domain which is at least four times as large as the domain described above, because only partial use of symmetry can be made. In such problems adequate wave damping along the walls has to be provided as well. This survey indicates that the hydro-dynamic coefficients of symmetrical bodies can well be determined by the finite element approach. However, the numerical solution of general three-dimensional motion problems is not yet feasible.

12.4.6 A moored ship close to a quay

This problem has been chosen to illustrate some of the complications the proposed method can deal with. The two-dimensional model drawn in Figure 12.6 shows a ship moored in shallow water. The mass of the ship per

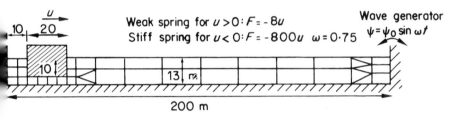

Figure 12.6 Determination of response of moored vessel

unit length is 2×10^5 kg/m. The draught of the ship is 10 m and its keel clearance is 3 m. The mooring is reflected by a bilinear spring whose compressive stiffness is much greater than that in tension. The spring coefficient (c) per unit length and the corresponding frequencies (excluding the influences of added mass of the water), calculated as $\omega = \sqrt{c/m}$, are

$$\text{tension} \quad : c_t = 8 \, \text{kN/m}^2 \quad \Rightarrow \omega_t = 0.2 \, \text{rad/s}$$

$$\text{compression}: c_c = 800 \, \text{kN/m}^2 \Rightarrow \omega_c = 2 \, \text{rad/s}$$

The frequency of the wave generator ($\omega = 0.75$ rad/s) lies between these values. The waves travel at a speed corresponding with their group velocity in shallow water from the wave generator to the model.[2] The waves will be reflected by the quay which increases the local wave heights. The experiment in this model was terminated after the waves, reflected successively both by the model and the wave generator, reappeared at the model. This time can be calculated from the group velocity. Figure 12.7 shows the response of the floating body with and without mooring in terms of its horizontal displacement. In the given time interval the maximum mooring force is found during compression of the stiff spring for $t \approx 30$ s.

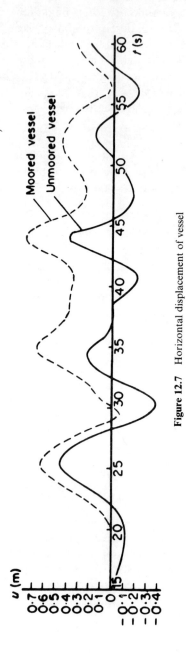

Figure 12.7   Horizontal displacement of vessel

12.5 Conclusions

(1) The coupled problem of a floating structure in an ideal, non-viscous fluid can be solved with the finite element method. In this way more complicated boundaries of the body and changes in water depth can be dealt with.

(2) Two-dimensional ship motion problems can be solved efficiently with the method. However, owing to limitations in computing, this numerical approach can be applied to simple three-dimensional problems only.

(3) The elements at the free surface should be less than or equal to a quarter of the minimum wave length. The maximum time step Δt in the fourth-order Runge–Kutta method is equal to the square root of the minimum element length divided by the gravity acceleration.

(4) The application of very fine meshes of elements is very costly and the corresponding improvement of results is marginal in comparison with the distribution used in this paper.

References

1. J. J. Stoker, *Water Waves*, Interscience, New York, 1957.
2. G. Vossers, *Behaviour of Ships in Waves*, Stam, Haarlem, 1962.
3. J. P. Hooft, 'Hydrodynamic aspects of semi-submersible platforms', *Ph.D. Thesis*, Technological University, Delft, 1970.
4. O. C. Zienkiewicz, *The Finite Element Method in Engineering Science*, McGraw-Hill, London, 1971.
5. W. Visser, 'The application of the finite element method to deformation and heat conduction problems', *Ph.D. Thesis*, Technological University, Delft, 1968.
6. J. D. Opsteegh, 'The prediction of regular wave forces on floating bodies with the finite element method', (in Dutch), *Report Appl. Math., Techn. Univ. Delft*, 1971.
7. P. C. Chowdhury, 'Fluid finite elements for added mass calculations', *Int. Shipbuilding Progress*, **19**, 1972.
8. A. R. Mitchell, 'Variation principles—a survey', *NATO advanced study institute at Kjeller (Norway) on Numerical Solutions of Partial Differential Equations, August 1973*.
9. B. A. Finlayson, *The Method of Weighted Residuals and Variational Principles*, Academic Press, New York, 1972.
10. G. Strang and G. J. Fix, *An Analysis of the Finite Element Method*, Prentice-Hall, Englewood Cliffs, N.J., 1973.
11. J. H. Vught, 'The hydrodynamic forces and ship motions in waves, *Ph.D. Thesis*, Technological University, Delft, 1970.

Chapter 13

Linear Wave Propagation Problems and The Finite Element Method

J. C. W. Berkhoff

13.1 Introduction

Computational methods for wave propagation problems are of great interest for the designers of harbours, offshore structures and coastal protection works. In general, several phenomena such as breaking of the waves, physical irregularity, viscosity and energy dissipation play a part in water wave propagation problems. A mathematical model with all of these effects taken into account would be too complicated so the mathematical model described in this paper will be restricted to linear, simple harmonic, irrotational water waves.

In the linear theory of regular waves there are three kinds of problems:

(1) The diffraction problem—waves propagating over a horizontal bottom are disturbed by the presence of an obstacle of arbitrary shape (wave penetration into a harbour; wave forces upon submerged structures).
(2) The refraction problem—an uneven bottom with moderate slopes changes the wave characteristics such as wave length, wave height and direction of propagation (wave ray theory).
(3) The combined refraction–diffraction problem—waves propagating over an uneven bottom are disturbed by the presence of an obstacle and by the variation of the bottom.

The diffraction problem can be treated numerically by the well known discrete source distribution method,[1,5] which can be classified as a finite element method. The refraction problem can be solved with the aid of characteristic equations defining the path of a wave ray.[2] The computational method for the combined refraction–diffraction problem is a mixture of a source distribution method and a variational method, and therefore also suitable for the finite element technique.[3]

13.2 Equations

13.2.1 Basic equations

The mathematical model is restricted to linear simple harmonic and irrotational water waves, so the basic equations for the model are:[6,7]

Field equation: $$\nabla^2 \phi = 0 \qquad (13.1)$$

Free surface equation: $\dfrac{\partial \phi}{\partial z} - \dfrac{\omega^2}{g}\phi = 0$ at $z = 0$ $\hspace{1cm}$ (13.2)

Bottom condition:

$$\frac{\partial \phi}{\partial x}\frac{\partial h'}{\partial x} + \frac{\partial \phi}{\partial y}\frac{\partial h'}{\partial y} + \frac{\partial \phi}{\partial z} = 0 \quad \text{at } z = -h'(x, y) \hspace{1cm} (13.3)$$

where ϕ (x, y, z) is the three-dimensional complex wave potential, ϕ_1 and ϕ_2 are the real and imaginary parts of ϕ, $i = \sqrt{-1}$, ∇^2 is the Laplace operator, g is the acceleration due to gravity, ω is angular frequency, x and y are the horizontal coordinates and z is the vertical coordinate. Also $h'(x, y) = (H + h)(x, y)$, where h is the free surface elevation measured from the mean water level and H is the water depth to the bottom, again measured from the mean water level.

Once the wave potential function ϕ has been found, all physical quantities of interest can be computed, for instance:

The free surface elevation:

$$h(x, y, t) = \frac{\omega}{g}(\phi_1 \sin \omega t - \phi_2 \cos \omega t) \quad \text{at } z = 0 \hspace{1cm} (13.4)$$

The linearized pressure:

$$p(x, y, z, t) = -\rho g z + \omega \rho(\phi_1 \sin \omega t - \phi_2 \cos \omega t) \hspace{1cm} (13.5)$$

The velocity vector

$$\mathbf{v}(x, y, z, t) = \nabla \phi_1 \cos \omega t + \nabla \phi_2 \sin \omega t \hspace{1cm} (13.6)$$

with

$$\nabla = \text{gradient operator } (\partial/\partial x, \partial/\partial y, \partial/\partial z)$$

The wave height:

$$\bar{H}(x, y) = \frac{2\omega}{g}\sqrt{\phi_1^2 + \phi_2^2} \quad \text{at } z = 0 \hspace{1cm} (13.7)$$

The phase of the free surface elevation:

$$S(x, y) = -\arctan(\phi_1/\phi_2) \quad \text{at } z = 0 \hspace{1cm} (13.8)$$

The energy flux vector:

$$\mathbf{E} = \rho\omega[\phi_2\nabla\phi_1 \cos^2 \omega t - \phi_1\nabla\phi_2 \sin^2 \omega t + \tfrac{1}{2}(\phi_1\nabla\phi_2 - \phi_2\nabla\phi_2)\sin 2\omega t]$$

$$\hspace{1cm} (13.9)$$

13.2.2 *The reduced diffraction–refraction equation*

The derivation of the reduced diffraction–refraction equation starts with the basic equations (13.1) to (13.3) by introducing dimensionless coordinates with the aid of the characteristic length $l = g/\omega^2$ at the free surface and the mean waterdepth H. For the slope of the bottom, however, the horizontal length L will be used as a characteristic length (see Figure 13.1).

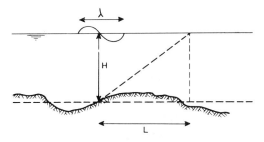

Figure 13.1   Characteristic lengths

Introduce $x' = x/l$, y'/l; $\zeta = z/H$ and $h'' = h'/H$; $\bar{x} = x/L$ and $\bar{y} = y/L$.

The equations written in these dimensionless quantities are, omitting primes for simplicity in notation:

$$\frac{\partial^2 \phi}{\partial x^2} + \frac{\partial^2 \phi}{\partial y^2} + \left(\frac{l}{H}\right)^2 \frac{\partial^2 \phi}{\partial \zeta^2} = 0 \tag{13.10}$$

$$\frac{\partial \phi}{\partial \zeta} - \frac{H}{l} \phi = 0 \quad \text{at } \zeta = 0 \tag{13.11}$$

$$\frac{\partial \phi}{\partial \zeta} + \frac{H^2}{lL}\left(\frac{\partial \phi}{\partial x}\frac{\partial h}{\partial \bar{x}} + \frac{\partial \phi}{\partial y}\frac{\partial h}{\partial \bar{y}}\right) = 0 \quad \text{at } \zeta = -h \tag{13.12}$$

Assuming the potential function ϕ can be written in the form

$$\phi(x, y, z) = Z\left(h, \zeta ; \frac{H}{l}\right)\varphi\left(x, y, \frac{H}{\sqrt{lL}}, \zeta\right) \tag{13.13}$$

and developing the function φ into a Taylor-series with respect to $(H/\sqrt{lL})\zeta$

$$\varphi\left(x, y, \frac{H}{\sqrt{lL}}, \zeta\right) = \varphi_0(x, y) + \frac{H}{\sqrt{lL}}\zeta\varphi_1(x, y) + \frac{H^2}{lL}\zeta^2\varphi_2(x, y) + \cdots \tag{13.14}$$

the following results with dimensional quantities can be obtained:[3]

$$Z(h, z) = \frac{\cosh\{n_\omega(h + z)\}}{\cosh(n_\omega h)} \tag{13.15}$$

and the equation

$$\frac{\partial}{\partial x}\left(\frac{\omega v_g}{n_\omega}\frac{\partial \varphi_0}{\partial x}\right) + \frac{\partial}{\partial y}\left(\frac{\omega v_g}{n_\omega}\frac{\partial \varphi_0}{\partial y}\right) + \omega n_\omega v_g \varphi_0 = 0 \tag{13.16}$$

where n_ω is the wave number, the positive real root of the dispersion relation; $\omega^2 = gn_\omega \tanh(n_\omega h')$; and v_g is the group velocity

$$= \frac{1}{2}\left(1 + \frac{2n_\omega h}{\sinh 2n_\omega h}\right)\frac{\omega}{n_\omega}$$

For deep water or for water with a constant water depth Equation 13.16 becomes the well known Helmholtz equation, describing the diffraction problem. For shallow water Equation 13.16 becomes the linearized shallow water equation.

13.3   Method of solution

13.3.1   Variational formulation and source distribution

The solution of the differential Equation 13.16 over an arbitrary area can be found by minimizing the corresponding functional over this area, taking into account the conditions at the boundaries, i.e. full reflection at rigid walls and the radiation condition at infinity. To overcome the difficulty of the condition at infinity the area is divided into two subareas (see Figure 13.2). Area I

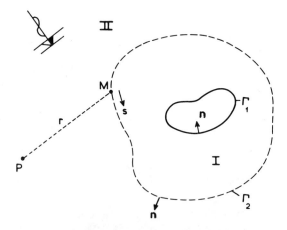

Figure 13.2   Definition sketch of solution domain

corresponds to the area where the influence of the variation of the bottom can be expected to be important. In area II the water is deep or the water depth is a constant.

The solution in area II will be a superposition of an incident wave and an outgoing wave which is caused by the presence of an obstacle or by the variation of the bottom in area I. This outgoing wave will be represented in its turn by a superposition of waves coming from sources at the boundary between area I and area II and satisfying the radiation condition at infinity.

The solution at this boundary Γ must be continuous with respect to wave height and phase, giving the additional equations to determine the intensity of the source distribution.

The functional, which must be minimized to get the solution of Equation 13.16 in area I with a variable water depth, reads, omitting the subscript zero.[4]

$$\Pi = \frac{1}{2} \int \int_{I} \left[\frac{\omega v_g}{n_\omega} (\nabla \varphi \cdot \overline{\nabla \varphi}) - \omega n_\omega v_g \varphi \overline{\varphi} \right] dx \, dy \qquad (13.17)$$

with the overbar denoting the conjugate complex value and ∇ now the two-dimensional gradient operator. Minimizing Equation 13.17 gives a solution with the natural boundary conditions:

$$\frac{\partial \varphi}{\partial \mathbf{n}} = 0 \quad \text{at } \Gamma_1 \text{ and } \Gamma_2$$

with $\mathbf{n}$ the normal vector of the boundary.

If it is assumed, that the boundary condition at Γ_2 is

$$\frac{\partial \varphi}{\partial \mathbf{n}} = f$$

with an arbitrary function f, then the following terms must be added to the functional Π:

$$-\frac{1}{2} \int_{\Gamma_2} (f \overline{\varphi} + \overline{f} \varphi) \frac{\omega v_g}{n_\omega} d\Gamma$$

In area II the solution must satisfy the Helmholtz equation of the diffraction problem and can be written in the form[1]

$$\psi(P) = \tilde{\varphi}(P) + \int_{\Gamma_2} q(s) \frac{1}{2i} H_0^1(n_\omega r) \, d\Gamma \qquad (13.18)$$

corresponding to the source distribution theory, with

$\psi(P) = $ the solution in a point P of area II
$\tilde{\varphi}(P) = $ the potential in P corresponding to the known incident wave field
$q(s) = $ the source intensity function along the boundary Γ_2
$H_0^1 = $ Hankel function of the first kind and zeroth order, satisfying the Helmholtz equation and the Sommerfeld condition
$r = $ the distance between point P and the current point along the boundary.

The source intensity function q along the boundary Γ_2 must satisfy the integral equation:

$$\left(\frac{\partial \psi}{\partial \mathbf{n}}\right)_P = \left(\frac{\partial \tilde{\varphi}}{\partial \mathbf{n}}\right)_P + q(P) + \int_{\Gamma_2} q(s)\frac{\partial}{\partial \mathbf{n}}\left[\frac{1}{2i}H_0^1(n_\omega r)\right]d\Gamma \qquad (13.19)$$

with P situated on the boundary Γ_2.

With the two continuity conditions

$$\psi = \varphi \qquad \text{and} \qquad \frac{\partial \psi}{\partial \mathbf{n}} = \frac{\partial \varphi}{\partial \mathbf{n}}(=f) \qquad (13.20)$$

at the boundary Γ_2 the problem is well defined and the unknown functions q and φ can be found. Once the intensity of the source distribution has been computed, the solution ψ in area II can be obtained in any desired point P, using Equation 13.18.

13.3.2   Numerical treatment

The numerical approach to find the minimum of the functional, Equation 13.17, and the intensity of the source distribution along Γ_2, is based on the finite element method.[8] Splitting up the area I into N elements of triangular form and approximating both the real and imaginary parts of the potential function φ in each element by a linear expression with respect to the variables x and y gives:

$$\varphi^e(x, y) = \mathbf{N}^e(x, y) \cdot \boldsymbol{\varphi}^e \qquad (13.21)$$

in which $\boldsymbol{\varphi}^e = $ the vector of values of φ in the angular points of the eth element and $\mathbf{N}^e = $ the vector of coefficients, which are functions of x and y.

The coefficients are given by (see Figure 13.3)

$$\mathbf{N}^e(x, y) = \begin{pmatrix} a_l^e & b_l^e & c_l^e \\ a_j^e & b_j^e & c_j^e \\ a_k^e & b_k^e & c_k^e \end{pmatrix}\begin{pmatrix} 1 \\ x \\ y \end{pmatrix} \qquad (13.22)$$

in which

$$a_l^e = -\frac{1}{2\Delta_e}(y_j^e x_k^e - x_j^e y_k^e)$$

$$b_l^e = \frac{1}{2\Delta_e}(y_j^e - y_k^e) \qquad \text{and} \qquad c_l^e = -\frac{1}{2\Delta_e}(x_j^e - x_k^e)$$

and $a_j^e, a_k^e, b_j^e, b_k^e$ and c_j^e, c_k^e can be found by changing l, j and k in a cyclic order.

Substitution of the linear approximation (13.22) into the expression of the functional gives a quadratic function in the M nodal values of the potential

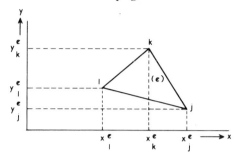

Figure 13.3 Triangular element

function φ for the total number of nodal points in area I:

$$\Pi = \sum_{e=1}^{N} \frac{1}{2} \int \int_{\Delta_e} \left[\frac{\omega v_g}{n_\omega} \{ \nabla(\mathbf{N}^e . \boldsymbol{\varphi}^e) . \overline{\nabla(\mathbf{N}^e . \boldsymbol{\varphi}^e)} - n_\omega^2 (\mathbf{N}^e . \boldsymbol{\varphi}^e)\overline{(\mathbf{N}^e . \boldsymbol{\varphi}^e)} \} \right] dx\, dy$$

$$- \sum_{m=1}^{R} \int_{L_m} \left[\frac{\omega v_g}{n_\omega} \{ (\mathbf{N}^m . \boldsymbol{\varphi}^m) \bar{f} + \overline{(\mathbf{N}^m . \boldsymbol{\varphi}^m)} f \} \right] d\Gamma \qquad (13.23)$$

with L_m = the segment of the boundary Γ_2 belonging to the mth element and R = the number of elements with one side coinciding with the boundary Γ_2.

In the same way the source intensity function q in each segment is approximated by the expression:

$$q(s) = \mathbf{M}^m(s) . \mathbf{q}^m \qquad (13.24)$$

with $\mathbf{q}^m = $ the vector of values of q in the computing points of the mth segment of the boundary Γ_2 and $\mathbf{M}^m = $ the vector of coefficients, which are functions of the integration variable Γ.

Taking only one computing point in each segment (see Figure 13.4) and the approximation function identically one, the integral (13.19) written in discrete form is then:

$$f(P) = \left(\frac{\partial \tilde{\varphi}}{\partial n} \right)_P + q_P + \sum_{m=1}^{R} q_m \frac{\partial}{\partial \mathbf{n}_P} \left[\frac{1}{2i} H_0^1(n_\omega r_{PP_m}) \right] L_m \qquad (13.25)$$

and the value of the potential function in the computing point P:

$$\varphi(P) = \tilde{\varphi}(P) + \sum_{m=1}^{R} q_m \frac{1}{2i} H_0^1(n_\omega r_{PP_m}) L_m \qquad (13.26)$$

The expression (13.25) for the function f is substituted into the functional (13.23) which must be minimal with respect to variations in the M nodal values of both the real and imaginary parts of the potential function φ, giving twice a set of M linear equations for these nodal values and for the unknown source intensity function in the R computing points along the boundary Γ_2. Taking the value of φ in the source point P as the average of the values in the neighbouring nodal points P_i and P_j (see Figure 13.4):

$$\varphi(P) = \tfrac{1}{2}\{\varphi(P_i) + \varphi(P_j)\} \tag{13.27}$$

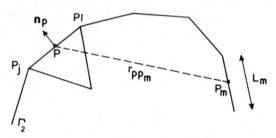

Figure 13.4 Polygon approximation

then Equation 13.26 supplies a set of R linear equations for the complex values of the potential function φ in the nodal points and the intensity function q in the computing points along the boundary Γ_2.

In matrix form the following complex set of equations must be solved:

$$\begin{pmatrix} A & B \\ C & D \end{pmatrix} \begin{pmatrix} \varphi \\ q \end{pmatrix} = \begin{pmatrix} r_1 \\ r_2 \end{pmatrix} \tag{13.28}$$

Thus

$$\begin{aligned} A\varphi + Bq &= r_1 \\ C\varphi + Dq &= r_2 \end{aligned} \tag{13.29}$$

with A = a real symmetric $M \times M$ matrix with a band structure generated by the finite element method,

B = a complex $M \times R$ matrix with non-zero values in the rows that correspond with the nodal points along the boundary Γ_2,

C = a real $R \times M$ matrix generated by the averaging procedure (13.27),

D = a complex $R \times R$ matrix with coefficients consisting of Hankel functions according to Equation 13.25.

φ = the vector of the unknown complex values of φ in the M nodal points,

$\mathbf{q}$ = the vector of the unknown complex values of μ in the R computing points along the boundary Γ_2,

$\mathbf{r}_1$ = a known vector provided by the normal derivative of the incident wave potential,

$\mathbf{r}_2$ = a known vector provided by the incident wave potential $\tilde{\varphi}$.

This series of equations are solved by a direct method of solution. First the vector of intensity values along the boundary Γ_2 is computed according to

$$\mathbf{q} = (\mathbf{D} - \mathbf{C}\mathbf{A}^{-1}\mathbf{B})^{-1}(\mathbf{r}_2 - \mathbf{C}\mathbf{A}^{-1}\mathbf{r}_1) \qquad (13.30)$$

and then the vector of potential values in the nodal points of area I by:

$$\boldsymbol{\varphi} = \mathbf{A}^{-1}\mathbf{r}_1 - \mathbf{A}^{-1}B\mathbf{q} \qquad (13.31)$$

The inverse $\mathbf{A}^{-1}$ is not computed, only the L–R decomposition of $\mathbf{A}$, taking into account the symmetrical band structure of the matrix.

With respect to the accuracy of the method of solution, one has to deal with the accuracy of the discrete source distribution method, the accuracy of the finite element method and with the influence of both methods upon each other. An analytical approach to the aspect of the accuracy of the total method of solution is very difficult, because of the number of factors playing a part, such as:

(a) the magnitude of the elements,
(b) the orientation of the elements with respect to the direction of wave propagation,
(c) the rate of approximation of the unknown functions in each element or segment,
(d) the representation of the boundaries by polygons,
(e) the accuracy of the numerical method of integration,
(f) the accuracy of the method of solution used for solving the ultimate set of linear equations.

Experimental results give the impression that the total error in the wave height is a function of the number of computing points over a distance of one wave length in the direction of propagation (see Figure 13.5).

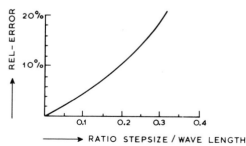

Figure 13.5   Relation between error and step size

13.4 Numerical results

To give an indication of the possibilities of the described model for two-dimensional wave propagation over an uneven bottom, an example is given for which the refraction model breaks down because of the existence of a caustic (see Figures 13.6 and 13.7). The results of the same problem, but now solved with the combined refraction–diffraction model, are given in Figures 13.8, 13.9 and 13.10. To save computing time and computer memory, only the most interesting part of the shoal has been taken into account. The refraction computations have given the values of the potential function at the boundary AA', assuming the results of the refraction model are fairly well up to there.

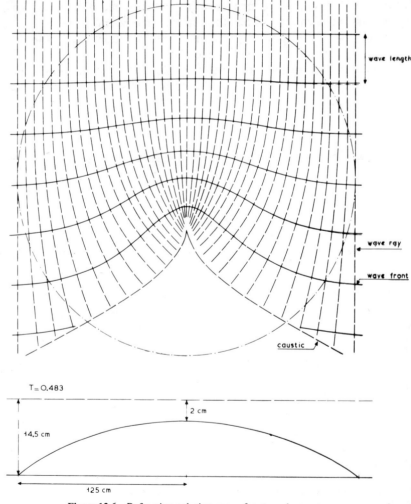

Figure 13.6   Refraction solution: wave fronts and wave rays

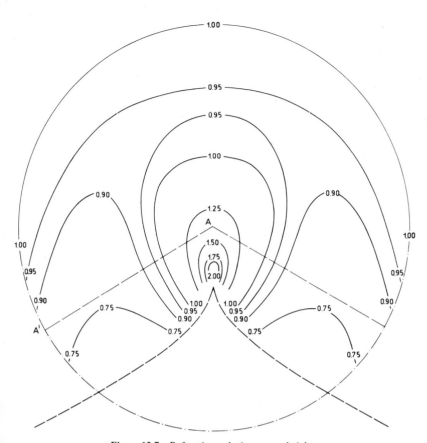

Figure 13.7 Refraction solution: wave heights

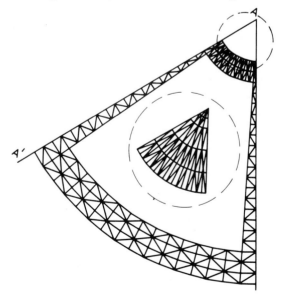

Figure 13.8 Refraction–diffraction solution: configuration of elements

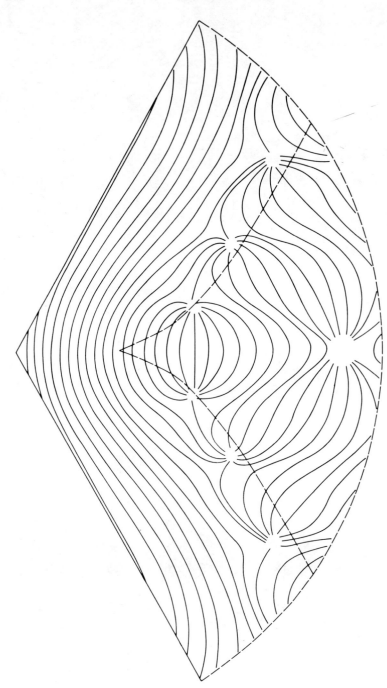

Figure 13.9 Refraction–diffraction solution: lines of equal phase lines every $\pi/4$ rad

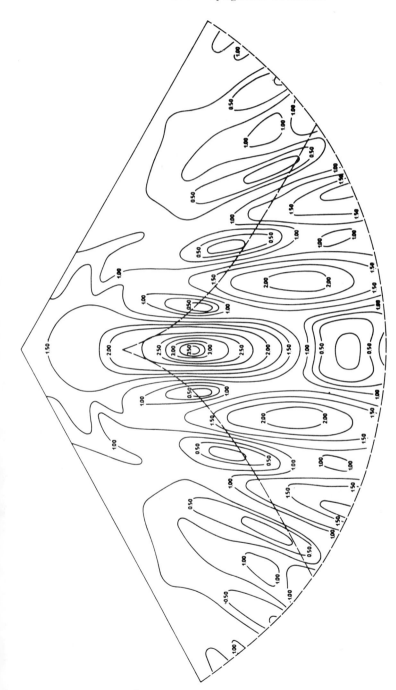

Figure 13.10 Refraction–diffraction solution: lines of equal wave height, lines every 0.25 unit

Finite Elements in Fluids

13.5 Conclusions

Mathematical models for linear simple harmonic waves can be of great help to give quantitative or qualitative information about the phenomenon of wave propagation. The main restriction of these models is the numerical requirement of having about five computing points over one wave length to compute the potential function with an accuracy of roughly five per cent. For the combined refraction-diffraction model this requirement is a rigid one with respect to computing time and computer memory in the case of a large area compared with the mean wave length. However, it is possible to divide the whole area of interest into subareas for which different models (pure diffraction, pure refraction or the combined model) can be applied separately and to join the different solutions in an appropriate way.

References

1. J. C. W. Berkhoff, 'Computation of the phenomenon of diffraction', Delft Hydraulics Laboratory, *Report on Mathematical Investigations, S5-1*, 1969 (in Dutch).
2. J. C. W. Berkhoff, 'Wave refraction: derivation and numerical solution of the refraction equations', Delft Hydraulics Laboratory, *Report on Mathematical Investigation, S5-11*, 1970.
3. J. C. W. Berkhoff, 'Refraction and diffraction of water waves: derivation and method of solution of the two-dimensional refraction-diffraction equation', Delft Hydraulics Laboratory, *Report and Mathematical Investigation, W 154*, 1973.
4. R. Courant and D. Hilbert, *Methods of Mathematical Physics*, Vol. 2, Interscience, New York, 1962.
5. C. J. Garrison and P. Y. Chow, 'Wave forces on submerged bodies', *Proc. Am. Soc. Civ. Eng.*, **98**, WW3, 375–392 (1972).
6. H. Lamb, *Hydrodynamics*, sixth ed., Cambridge University Press, 1963.
7. J. J. Stoker, *Water Waves*, Interscience, New York, 1957.
8. C. Taylor, B. S. Patil and O. C. Zienkiewicz, 'Harbour oscillations: a numerical treatment for undamped natural nodes', *Proc. Inst. Civ. Engrs.*, 43, 141–155 (1969).

Chapter 14

Finite Element Analysis of Jet Impingement on Axisymmetric Curved Deflectors

T. Sarpkaya and G. Hiriart

14.1 Introduction

The deflection of a free jet by a solid boundary is well suited to potential-flow analysis because of the dominance of inertia and pressure intensity in the establishment of the flow pattern. The design of impulse machinery, thrust reversers, flip buckets on spillways, etc., utilizing this momentum change could be facilitated greatly if the idealized geometry of the system under potential flow conditions were known, because such conditions represent asymptotic values which are approached as the effects of secondary variables such as entrainment, boundary layer, compressibility, jet attachment to adjacent surfaces (Coanda effect) etc. are decreased. With such information available, refinements of design could be based upon a secure knowledge of the fundamentals and many rules of thumb could be replaced with precise quantitative data. Specifically, if the total angle through which the jet is deflected is determined for conditions of both partial and complete interception by the boundary, then the principle of impulse and momentum can be used to compute forces or other dynamic characteristics of the system.

The two-dimensional counterpart of the jet deflection problem has been treated by several investigators through the use of the powerful analytic-function theory and successive conformal transformations. Sarpkaya[1] solved the U-shaped, two-segment, deflector problem where the turning angle between the segments is limited to 90 degrees. Tinney and coworkers[2] extended this analysis to the case where the turning angle between the symmetrically situated segments is greater than 90 degrees. Later, Chang and Conly[3] presented an analysis for a bucket composed of a series of segments of arbitrary number, lengths and angles. However, the basic as well as practical problem of the direct and exact analysis of jet deflection by curved buckets remained unsolved. Recently, Sarpkaya and Hiriart[4] presented a novel solution for the deflection of a two-dimensional jet by a finite curved bucket through the use of the singular integral equations of Riemann and Hilbert, the modified-hodograph method and the generalized free streamline analysis of Levi-Civita. The fact should be emphasized that from the mathematical viewpoint the direct calculation of the free streamline flows is restricted to two-dimensional irrotational flows which are free from gravitational effects. Indirect methods employing suitable functions which satisfy the boundary conditions dictated by the gravitational forces are applicable only to simplest geometries. Thus, recourse is usually made to approximate methods of analysis where a free surface is assumed and its suitability is assessed from approximate potential solutions obtained through the use of various numerical techniques.

The three-dimensional counterpart of the jet deflection problem has not been solved in any generality. Attempts to formulate an exact solution have been mostly unsuccessful even for axisymmetric inviscid flows with no body forces. The case of a circular jet striking a plate normally was analysed by Schach[5] using approximate methods similar to those of Trefftz[6] with successive adjustment of the free streamline. Jeppson[7] applied the finite difference technique to the solution of two, axisymmetric, potential-flow problems, namely to that of flow from a nozzle and of the cavitating flow of a jet past a body of revolution. Other noteworthy contributions to the analysis of the jet efflux from nozzles and orifices were made by Southwell and Vaisey,[8] Rouse and Abul-Fetouh,[9] Garabedian[10] and Hunt[11] through the use of the relaxation and finite difference methods. Schnurr and coworkers[12] used the relaxation method to analyse the turning of two-dimensional and axisymmetric jets from curved surfaces where there was only one free-stream surface, i.e. the jet was assumed to leave the deflector exactly parallel to the tangent at the lip of the deflector surface and the consequences of the difference between the actual deflection angle and the said tangent to the deflector surface was taken as a measure of 'spillage' and expressed in terms of a turning-effectiveness coefficient determined experimentally. Thus, their analysis does not constitute a solution to the problem under consideration. Notwithstanding this fact, their method, as well as those used by others, suffer from convergence and accuracy problems and merely resort to simple trial-and-error procedures to locate the free surface. Suffice it to say that a method was needed which could yield solutions of a prescribed accuracy for a wide variety of fluid flow problems involving Dirichlet, Neumann and mixed boundary conditions.

Zienkiewicz and Cheung[13] proposed in 1965 the application of the finite element method to the solution of field problems involving the equations of Laplace and Poisson. Since then a significant number of applications of the finite element method to fluid dynamics has appeared in the literature (see for example Norrie and de Vries[14] and the references cited therein). The method has recently been applied to several jet efflux problems involving only one free stream surface and relatively small jet contraction by Chan and Larock.[15] Suffice it to say that the finite element method has proved its versatility and applicability to a wide variety of two-dimensional and axisymmetric fluid flow problems. The truly three-dimensional problems such as the oblique impact of an axisymmetric jet on a plane surface, to name just one, offer considerable challenges and demand additional sophistication.

The present study, encouraged by the success of the finite element method, is devoted to a determination of the angle of deflection, the location of the free stream surfaces, and the velocity and pressure distributions caused by a finite hemispherical boundary placed symmetrically with respect to the axis of an axisymmetric, inviscid jet issuing from a nozzle (see Figure 14.1). This problem has not previously been analysed by either exact or approximate methods.

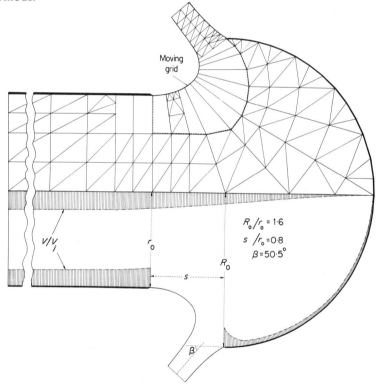

Figure 14.1 Finite element grid for an axisymmetric jet impinging upon a hemispherical thrust reverser (Case no. 4 in Table 14.1)

14.2 Analysis

14.2.1 *Governing equation and functional*

The fluid is assumed to be inviscid, incompressible and free from body forces. Consequently, the problem may be formulated in terms of either the velocity potential ϕ or the Stokes stream function ψ. Here the potential function is chosen as the unknown field parameter. The governing equation is then given by

$$\phi_{,xx} + \phi_{,r}/r + \phi_{,rr} = 0 \tag{14.1}$$

For an axisymmetric flow, the solution to the Laplace field equation satisfying the specified normal-velocity boundary conditions $(\phi_{,n})^a$ is given by that admissible function ϕ which minimizes the function (see Reference 14).

$$I(\phi) = \pi\rho \iint_{\Omega} [(\phi_{,x})^2 + (\phi_{,r})^2] r \, dr \, dx - 2\pi\rho \int_{\Gamma} \phi(\phi_{,n})^a r \, d\Gamma \tag{14.2}$$

in which Ω is half of a meridional section of the flow and Γ is a portion of the curve bounding this area where the normal derivative is prescribed. The first integral in Equation 14.2 represents the kinetic energy of the fluid within the entire control volume and the second integral represents twice the work done by the impulsive pressure $\rho\phi$ on the boundaries in starting the motion from rest.

14.2.2 Finite element representation

Finite elements of triangular shape were employed in the present work. In choosing the order of approximation it was felt that a linear representation of ϕ, which gives constant velocities along any line within an element, results in unacceptably large errors in the velocity gradients between two adjacent elements. A second-order polynomial, however, allows at least a linear variation of the velocity components and was therefore chosen.

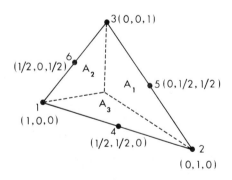

Figure 14.2 Triangular element and nodal points

The element is shown in Figure 14.2. The representation of ϕ can be written as

$$\phi = \mathbf{N}\boldsymbol{\phi}^e \tag{14.3}$$

where

$$\boldsymbol{\phi}^e = \lfloor \phi_1 \ \phi_2 \ \phi_3 \ \phi_4 \ \phi_5 \ \phi_6 \rfloor^{\mathrm{T}} \tag{14.4}$$

(the superscript e denotes that the vector refers to the individual element) and

$$\mathbf{N} = \lfloor N_1 \ N_2 \ N_3 \ N_4 \ N_5 \ N_6 \rfloor \tag{14.5}$$

the respective terms of $\mathbf{N}$, expressed as a function of triangular area coordinates, are given in Equation 4.20 in Chapter 4 of this volume. Substitutions of Equation 14.3 into the relevant functional (14.2) and application of the stationary condition of the functional results in the following set of element equations

$$\mathbf{K}^e \boldsymbol{\phi}^e = \mathbf{P}^e \tag{14.6}$$

where

$$\mathbf{K}^e = 2\pi\rho \int_{\Omega} (\mathbf{N}_{,x}^{\mathrm{T}} \mathbf{N}_{,x} + \mathbf{N}_{,r}^{\mathrm{T}} \mathbf{N}_{,r}) r \, dr \, dx \tag{14.7}$$

$$\mathbf{P}^e = 2\pi\rho \int_{\Gamma^e} (\phi_{,n})^a \mathbf{N} r \, d\Gamma^e \tag{14.8}$$

and $\mathbf{N}_{,x}$ and $\mathbf{N}_{,r}$ represent vectors of derivatives of $\mathbf{N}$ with respect to x and r, respectively. Γ^e is the portion of the element boundary subject to the condition $(\phi_{,n})^a$.

It follows that the global equations can be written in the form

$$\mathbf{K}\boldsymbol{\phi} = \mathbf{P} \tag{14.9}$$

where $\mathbf{K}$ and $\mathbf{P}$ are the global matrices obtained after appropriate summation of the corresponding element matrices and $\boldsymbol{\phi}$ is a column vector.

14.2.3 Iteration scheme

As cited in the introduction, the deflected jet is characterized by two, initially known, axisymmetric free stream surfaces. Thus special attention must be focused on finding a suitable iterative procedure for systematically approaching the final positions of the free boundaries. The iteration is terminated when the boundary conditions are satisfied within a prescribed absolute maximum error. The boundary condition to be satisfied is that the free surfaces be streamlines of constant velocity, that is, that the velocity normal to the free surface be zero and that the pressure on the surface be constant.

At various stages in the work the iteration scheme has been altered in detail. Here only the final scheme which has accomplished the foregoing objectives is presented.

Finite Elements in Fluids

14.2.4   Construction of the finite element grid

The grid is divided into two major regions. The first region consists of elements whose coordinates are fixed once and for all (see Figures 14.1 and 14.4). This region is well within the interior of the flow, i.e. near the

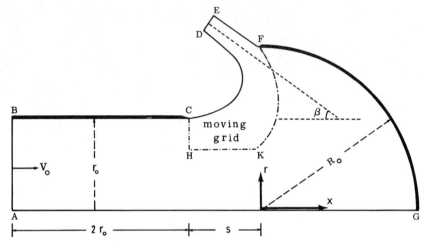

Figure 14.3   Geometry of the nozzle-reverser combination

hemispherical bucket, and is not expected, in the final solution, to be inter-sected by the free stream surfaces. Otherwise, a new fixed-grid region has to be defined. This is not a serious demand on the part of the analysis since what is needed is no more than a brief exposure to the gross characteristics of two- or three-dimensional jets.

The second region, i.e. the moving grid, consists of elements whose coordinates are relocated each time the free surface is moved. The movement of the grid is made in such a manner that each element maintains a shape more or less compatible to its original shape. The procedure may best be described through the use of Figure 14.4. At each iteration, the free surface ABC is moved inward or outward, in a manner to be described later, and the coordinates of the nodes along DK and EH are recalculated such that AD = 0·15 AF and AE = 0·50 AF. Obviously, the coefficients 0·15 and 0·50 are somewhat arbitrary. Needless to say, the use of different coefficients do not alter the final result as long as the ratio of the lengths of the three sides of each element are kept close to unity within reasonable limits, i.e. A1 ≃ AD, DM ≃ DE, etc. The grid inside the exiting jet (along EF in Figure 14.3) is relocated by moving each *middle node* to its correct position between two opposite nodes on each side of the jet.

The attention to one more detail is necessary in selecting both types of grids. It is clear that there are regions in the flow where the velocities and

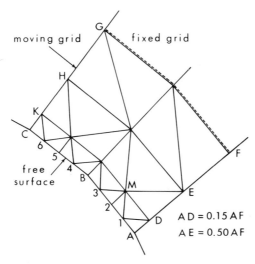

Figure 14.4 Definition of the moving grid along the free boundaries

their gradients are rather small (such as the bottom region of the bucket) and also there are regions where the velocities are higher their gradients are large (such as the free surfaces and regions near to the edge of the nozzle and the upper lip of the bucket). Thus in the regions of high velocity and/or velocity gradients, a grid composed of relatively small triangles must be used to ensure sufficient accuracy. In regions of rather low velocity gradients, however, larger elements may be used (see Figure 14.1). In the present analysis 288 elements were used. There were 683 nodal points and the bandwidth was 60.

14.2.5 Iteration of the free surfaces

The procedure will be described in several steps, each pertaining to a particular phase of the iteration scheme. Firstly, a deflection angle β is assumed and the two free surfaces are sketched (see Figure 14.3) as carefully as possible on the basis of past experience and familiarity with fluid dynamics. The assumed surfaces may differ considerably from their final shape provided that they do not, at some stage of the iteration process, intersect the fixed grid. Then the *assumed boundaries are regarded as rigid boundaries*, i.e. $\phi_{,n} = 0$ along the boundaries comprised of the free surfaces, the nozzle wall and the bucket. An arbitrary value (say 100) is assigned to ϕ at the nodal points along the boundaries comprised of the free surfaces, the nozzle wall and the Obviously, the results do not depend on the assumed value of ϕ.

The limits of the control volume are decided upon on the basis of the preliminary work with the problem under consideration. The primary consideration in the selection of the nozzle length BC and the deflected-jet length FE (see Figure 14.3) is that the velocity distribution across AB and DE must be nearly uniform. In the present analysis, the nozzle length was initially chosen to be $6r_0$. It has subsequently been found that a length of BC $= 2r_0$ will be more than sufficient in calculating V_0 to an accuracy of 0.5 per cent and will help to reduce significantly the number of the elements to be used. The jet length FE was varied from $0.6r_0$ to r_0 in various programs with no noticeable difference in the results.

Following the assumption of the deflection angle β and the jet length FE, the thickness of the jet DE $= d$, and the radial coordinates r_E and r_D were calculated from the equation of continuity by assuming $V_0 = 1$, (the velocity across AB). Obviously, d, r_E and r_D are subsequently recalculated on the basis of the iterated value of V_0. This will be explained further later. The point E is connected to the point F (the lip of the bucket) by a straight line as a first approximation to the upper free surface. Likewise, the point D is smoothly joined to the assumed lower free surface.

The foregoing provides all that is necessary for the initial data, i.e. the coordinates of the moving and fixed grid. The first run through the computer calculates the velocity at C, i.e. at the lip of the nozzle. Obviously, for a correct solution $V_C = 1.0$. However, since initially V_0 is assumed to be equal to unity, V_C turns out to be larger than unity. This is immediately corrected, to a first order of approximation, by writing

$$V_0 = V_0 - (V_C - 1.0)*V_0 \qquad (14.10)$$

This statement which is used during the subsequent iterations of the entire boundary, provides a new V_0 value. This, in turn, enables one to recalculate r_E, r_D and d. In passing it should be noted that the correction to r_E, r_D and d is usually very small and could be made even smaller by beginning the computation with a V_0 value somewhat smaller than unity (say $V_0 = 0.90$). Following this preliminary work, the ϕ values everywhere and the velocities along the assumed boundaries are calculated. Obviously, the velocities along CD and EF should everywhere be equal to unity. Since in general this will not turn out to be the case, the assumed stream lines must be adjusted. The procedure which has enabled the assumed boundaries to converge in a systematic manner to their final positions is described below.

Let the velocity at an arbitrary point along the assumed boundary be V_s. Then the boundary is moved inward, along a line normal or nearly normal to the boundary, if $V_s > 1$ and vice versa by an amount

$$\nabla V = (VS**2 - 1.0)*FAC \qquad (14.11)$$

in which FAC is an assigned multiplier. Experience has shown that the iteration will converge even for a very crudely assumed boundary for a value of FAC $\simeq 0.10$. In the present analysis FAC was taken equal to 0.015.

It is not difficult to devise programs where the value of FAC may be decreased from 0.10 to 0·01 as one converges the final boundary. As to the reason for using $(V_s^2 - 1·0)$ in the FORTRAN statement given by Equation 14.11, it was simply for the purpose of accentuating the error and accelerating the correction of the boundary points at which the velocities most differ from unity. Obviously, one could have just as well used $(V_s^3 - 1·0)$.

At the end of a given number of iterations, the velocities along one or both boundaries (calculated by considering only the contributions from those triangles having one side in common with the free surface) may approach values other than unity because the assumed deflection angle is not necessarily the correct one. Clearly, the deflection angle adjusts itself at the end of each iteration since r_E, r_D and d change and the free surfaces move inward or outward. The change in β per iteration was about 0·2 degrees. Thus, for a large correction in β, say 20 degrees, about 100 iterations were needed. Instead, five iterations were carried out for the assumed deflection angle and then β was incremented by 2 degrees, clockwise or counterclockwise depending on whether the velocities in the upper free surface were smaller or larger than unity. When the velocities everywhere were within $1 \pm 0·05$, β was no longer incremented and the number of iterations was increased to 25. The calculations were terminated when the velocities everywhere were within $1 \pm 0·015$. Then the final coordinates of the boundaries, the velocities along the solid boundaries and the ϕ values were recorded. The entire iteration for a given nozzle–bucket geometry required approximately 30 minutes on an IBM 360/67 computer. The program was written in FORTRAN and double-precision arithmetic was used.

At the end of each iteration, the corrected boundary was smoothed through the use of a generally available CURVE FIT subroutine by passing a curve through three major nodal points (A, B and C in Figure 14.4) and the coordinates of the intermediate points (1 through 6) were recalculated. The CURVE FIT matches the slopes at the points A and C to the slopes of the preceding and following curves and thereby provides a smoother boundary. It should be pointed out that there is no need for an exotic subroutine and that any suitable polynomial may be used to achieve the stated objectives.

The experiments with the overall program included, among other things, the checking of the degree of convergence of the iteration by purposely shifting the initial position of the deflected jet to a larger or smaller angle than that finally established. The correct free streamline positions were arrived at regardless of whether the analysis began with a larger or smaller initial deflection angle. Another important experiment conducted with the overall program was the use of two different FAC values, i.e. FACO = 0·001 and FACT = 0·020. After assuming the boundaries in the manner described above, FACO was used for the upper portion of the free jet and FACT for the lower portion, for about 10 iterations. Then the FAC values were interchanged. This procedure essentially amounted to first iterating on the lower boundary and then on the upper boundary and alternating the order of iteration until the boundary conditions were satisfied everywhere. This method was perfectly satisfactory but required a relatively larger number of iterations in arriving at the final result.

14.3 Examples and results

The method and the procedure described above have been used to analyse the characteristics of axisymmetric jet deflection from hemispherical target-type thrust reversers. As far as the finite element method is concerned, there is no special requirement that the reverser shape be hemispherical, i.e. it could be any axisymmetric shape.

Many uses for thrust reversers on jet aircraft have been proposed. They include braking the landing load, reversing or spoiling thrust during the landing approach so that 100 per cent engine speed may be maintained, and braking during diving manoeuvres to limit flight speed. To be used effectively, however, the reverser must give the desired amount of reverse thrust without affecting the engine operation. Also the design must lend itself to stowage with a minimum amount of toat-tail or base drag. A target-type hemispherical thrust reverser appears to satisfy these and other operational requirements. Furthermore, extensive data are available on the thrust modulation characteristics of this type of reverser for comparison with those predicted numerically.

The geometry of the nozzle–bucket combination is uniquely defined by the ratios R_0/r_0 and s/r_0 once the bucket and nozzle shape are decided upon. It is clear, at least from the laboratory experiments,[16] that there is a unique combination of R_0/r_0 and s/r_0 for which the reverse thrust is maximum. The determination of this combination and the calculation of the resulting thrust constitute the essence of the practical problem.

Solutions were obtained for a family of R_0/r_0 and s/r_0 values. The results obtained for each case included the positions of the free streamlines, the velocity-potential field, and the velocity distribution along the solid boundaries. The force acting on the bucket was calculated both through the integration of $p\,dA$, (dA the projected elemental area at the centre of which the pressure is p) and through the use of the momentum equation as applied to the entire control volume. The calculation of force acting on the bucket through the integration of $p\,dA$ offers no special problems.

The equation of momentum with reference to Figure 14.3 may be written as

$$\pi r_0^2 p_0 + \pi \rho r_0^2 V_0^2 + \pi \rho r_0 V_0 V_j \cos \beta = T \tag{14.12}$$

where T is the force necessary to hold the bucket at rest. The equation of Bernoulli written between B and a point along the free stream line yields

(see Figure 14.3)

$$p_0 + \rho V_0^2/2 = \rho V_j^2/2 \qquad (14.13)$$

where p_0 is the uniform pressure across AB, (above ambient), and ρ is the density of fluid. Combining Equations 14.12 and 14.13, one has

$$C_T = T/(\pi \rho r_0^2 V_j^2/2) = 1 + V_0^2/V_j^2 + 2(V_0/V_j)\cos \beta \qquad (14.14)$$

in which V_j is the velocity along the free streamline. In passing it should be noted that the error made in the calculation of the thrust coefficient C_T by assuming the jet leave the deflector exactly parallel to the tangent at the lip of the deflector, i.e. by writing $\beta = 0$, as was done by Schnurr and coworkers,[12] would be

$$\Delta C_T = 2(V_0/V_j)(1 - \cos \beta) \qquad (14.15)$$

This error could be rather large particularly for large angles of deflection and its correction through the use of another arbitrary parameter such as the 'spillage coefficient' introduced by Schnurr and coworkers is not justified. The foregoing discussion merely points out the fact that the correct calculation of the thrust requires the determination of the angle of deflection either in a manner similar to that described in the present paper or through the use of another numerical method.

In the aircraft industry, the efficiency of a thrust reverser is expressed in terms of a 'reverse-thrust ratio' η_R defined by

η_R = (actual reversed-jet thrust)/(forward jet thrust of the nozzle alone)

In other words, η_R is given by

$$\eta_R = (\pi \rho r_0^2 V_0 V_j \cos \beta)/(\pi \rho r_0^2 V_j^2) = (V_0/V_j)\cos \beta \qquad (14.16)$$

Evidently, $\eta_R = \eta_R(R_0/r_0, s/r_0)$ since both V_0 and β depend on R_0/r_0 and s/r_0 for the bucket and nozzle geometry under consideration. The results of the present analysis will be compared with those obtained experimentally in terms of η_R.

The computer analysis has been carried out for the geometrical configurations listed in Table 14.1. The results tabulated in Table 14.1 are also shown in graphical form in Figure 14.5. The coordinates of the computed free surfaces are given in Table 14.2 for a typical case, namely for the Case no. 4.

It is evident from Table 14.1, as well as from Figure 14.5, that the experimentally determined values of η_R are somewhat larger than those obtained numerically. There are several reasons for this difference, the two most important ones being the Coanda effect (see, for example Reference 17) and the nozzle pressure ratio. The Coanda effect, i.e. the tendency of a jet to attach to adjacent surfaces (in this case to the outer surface of the nozzle)

Table 14.1

Case no.	R_0/r_0	s/r_0	β (deg.) (theory)	β (exper.) (± 2 deg.)	η_R (theory)	η_R (exper.)* (Reference 16)	η_R (exper.)† (present)
1	1·90	0·10	18·5	10	0·76	0·81	0·78
2	1·80	0·40	20·3	13	0·77	0·82	0·80
3	1·70	0·60	46·8	39	0·65	0·81	0·74
4	1·60	0·80	50·5	47	0·52	0·74	0·60
5	1·50	0·90	61·1	57	0·39	0·61	0·44
6	1·80	0·80	35·2	31	0·73	—	0·75

* With a boat-tail nozzle and a nozzle–pressure ratio of 2·0.
† With a sharp-edged straight nozzle and a nozzle–pressure ratio of 2·0.

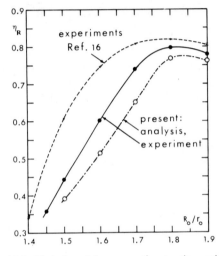

Figure 14.5 Variation of the reverse-thrust ratio η_R with R_0/r_0

because of the entrainment deprivation, decreases the deflection angle and thereby increases the reversed thrust. In the experiments conducted by Steffen and coworkers,[16] the outer surface of the nozzle was streamlined in the form of a boat-tail to decrease entrainment and thus to increase η_R. In fact, as noted by Steffen and coworkers, 'the pressure reductions on the boat-tail were large enough to account for as much as 20 per cent of the reverse–thrust ratio'. Thus, the relatively large differences between the η_R values experimentally obtained by Steffen and coworkers and those predicted numerically (see columns 6 and 7 in Table 14.1) are primarily attributable to the Coanda effect.

The present experiments were conducted with a straight, sharp-edged nozzle ($r_0 = 1·32$ to $1·57$ inches, $R_0 = 2·5$ inches) and the effect of the jet attachment has been minimized. This has resulted in a closer agreement

Table 4.2 Coordinates of the free stream lines for the Case no. 4 in Table 14.1 ($R_0/r_0 = 1.6$; $s/r_0 = 0.8$; $V_0 = 0.81$; $\beta = 50.5$ degrees; $\eta_R = 0.515$) (see Figure 14.3 for the coordinates x and r)

x/r_0	r/r_0	x/r_0	r/r_0	x/r_0	r/r_0
−0·800	1·000	−0·396	1·191	−0·332	1·337
−0·765	1·007	−0·388	1·200	−0·331	1·344
−0·729	1·015	−0·384	1·206	−0·331	1·350
−0·694	1·025	−0·379	1·212	−0·327	1·443
−0·660	1·036	−0·375	1·219	−0·367	1·513
−0·632	1·045	−0·371	1·225	−0·399	1·611
−0·603	1·055	−0·366	1·234	−0·495	1·742
−0·575	1·066	−0·361	1·242	−0·561	1·830
−0·548	1·078	−0·357	1·251	−0·627	1·918
−0·525	1·090	−0·353	1·260	−0·469	2·059
−0·503	1·102	−0·349	1·268	−0·397	1·977
−0·482	1·115	−0·346	1·276	−0·322	1·897
−0·461	1·130	−0·343	1·284	−0·255	1·809
−0·451	1·138	−0·341	1·293	−0·175	1·734
−0·440	1·147	−0·339	1·300	−0·158	1·675
−0·430	1·155	−0·337	1·308	−0·101	1·654
−0·420	1·165	−0·335	1·316	−0·090	1·588
−0·412	1·173	−0·334	1·325	0·000	1·600
−0·404	1·182	−0·333	1·331		

between the computed and experimental values of η_R (see columns 6 and 8 in Table 14.1). In passing it should be noted that the Coanda effect is not necessarily desirable, in spite of its contribution to the reverse thrust, for it may cause an unstable flow field and destructive vibrations.

The nozzle pressure ratio (total nozzle pressure/ambient pressure) or the actual velocity in the nozzle causes variations in η_R primarily because the entrainment needs of the deflected jet and hence the Coanda effect increase with increasing jet velocity. Consequently, η_R increases (about 10 percentage points) over a range of nozzle pressure ratios from 1·7 to 3·0. The experimental values of η_R tabulated in columns 7 and 8 in Table 14.1 were obtained with a pressure ratio of 2·0.

Finally, it is noted both from Figure 14.5 and Table 14.1 that the relative spacing of the nozzle and the reverser size significantly affect the flow reversal. Evidently, the reverse-thrust ratio increases with an increase in hemisphere diameter and reaches a maximum of about 80 per cent for $R_0/r_0 = 1.80$ and $s/r_0 = 0.40$. It is also evident from a comparison of the R_0/r_0, s/r_0 and η_R values for the second and sixth cases presented in Table 14.1 that there is an optimum nozzle–reverser spacing. Extensive experimental data obtained by Steffen and coworkers[16,18] with various types of reversers clearly show that at spacings greater than that required for optimum performance, the rever–thrust ratio may drop as much as 20 percentage points. Closer spacings do not noticably affect the reverser performance but result in decreased flow rate for a given total nozzle pressure or in increased pressure for a given flow rate because of the increased blockage or back-pressuring effect of the reverser.

Suffice it to say that the analysis of thrust reversers is extremely complex for the cases where the jet impinges asymmetrically or where the gravitational effects are important. Furthermore, some of the practical problems associated with thrust reversal, such as the reattachment of the jet to the nacelle of the engine, hot gas re-ingestion, interaction of the deflected jet with the ambient stream and the adjacent surfaces (Coanda effect), etc. cannot yet be analysed by any one of the existing numerical methods. Nevertheless, approximate analyses such as the one presented herein help to isolate the more promising types of thrust reversers, investigate their potentialities, evaluate their ideal performance characteristics, and to delineate the range of importance of the geometrical variables involved. Then those cases which are shown to be promising through computer experiments can be tested in a laboratory, at significantly reduced expense, to evaluate the effect of additional fluid-mechanical phenomena on their actual performance.

14.4 Conclusions

The work described herein has served two purposes. Firstly, it has demonstrated that the finite element method may be used with confidence for the analysis of an axisymmetric Laplace field where one or more parts of the boundary are to be determined as part of the solution. Secondly, it has yielded new engineering information concerning the ideal thrust-reversal characteristics of hemispherical cups. Such information is essential both in optimizing the performance characteristics of hemispherical reversers and in assessing the degree of significance of the additional real fluid effects such as the Coanda effect, viscous dissipation, pressure ratio, etc. on the actual performance of the optimized geometry.

The method and the procedure described herein may be used not only to analyse the effect of the various modifications to the hemispherical reverser (e.g. placing flat plates at various depths in the hemisphere, changing the lip angle, etc.) but also to analyse jet deflection from other physically important two-dimensional and axisymmetric thrust reversers.

References

1. T. Sarpkaya, 'Deflection of jets, II. Symmetrically placed U-shaped obstacle', *Free-Streamline Analysis of Transition Flow and Jet Deflection*, J. S. McNown and C.-S. Yih (eds.), State University of Iowa, Bulletin No. 35, 45–53, 1953.
2. E. R. Tinney, W. E. Barnes, O. W. Rechard and G. R. Ingram, 'Free-streamline theory for segmental jet deflectors', *Proc. ASCE, J. of the Hydraulics Division*, **87**, HY5, 135–145 (1961).

3. H. Y. Chang and J. F. Conly, 'Potential flow of segmental jet deflectors', *Journal of Fluid Mechanics*, **46**, 465–475 (1971).
4. T. Sarpkaya and G. Hiriart, 'Potential flow of curved jet deflectors', *Proceedings of the Canadian Congress of Applied Mechanics, Ecole Polytechnique, Montreal,* 665–666, 1973.
5. W. Schach, 'Umlenkung einer Kreisformigen Flussigkeitsstrahles an einer Ebenen Platte Senkrecht zur Stromungsrichtung', (Deflection of a circular jet by a plane plate perpendicular to the flow direction), *Ingenieur-Archiv*, **6**, 51–59 (1935).
6. E. Trefftz, 'Uber die Kontraction Kreisformiger Flussigkeitsstrahlen', (On the contraction of circular fluid jets), *Zeits. fur Math. und Physik*, **64**, 34–61 (1916).
7. R. W. Jeppson, 'Numerical solutions to free-surface axisymmetric flows', *Proc. ASCE, J. of the Engrg. Mech. Div.*, **95**, EM1, 1–20 (1969).
8. R. Southwell and G. Vaisey, 'Relaxation methods applied to engineering problems XII. Fluid motions characterized by free streamlines', *Phil. Trans. Roy. Soc. London, Ser. A.*, **240**, 117–161 (1948).
9. H. Rouse and A. Abul-Fetouh, 'Characteristics of irrotational flow through axially symmetric orifices', *Jour. of Applied Mechanics*, **17**, 4, 421–426 (1950).
10. P. Garabedian, 'Calculation of axially symmetric cavities and jets', *Pacific Jour. Math.*, **6**, 611–684 (1956).
11. B. W. Hunt, 'Numerical solution of an integral equation for flow from a circular orifice', *Journal of Fluid Mechanics*, **31**, 2, 361–377 (1968).
12. N. M. Schnurr, J. W. Williamson and J. W. Tatom, 'An analytical investigation of the impingement of jets on curved deflectors', *AIAA J.*, **10**, 11, 1430–1435 (1972).
13. O. C. Zienkiewicz and Y. K. Cheung, 'Finite elements in the solution of field problems', *The Engineer, London*, **24**, 507–510, Sept. 1965.
14. D. H. Norrie and G. de Vries, *The Finite Element Method—Fundamentals and Applications*, Academic Press, N.Y., 1973.
15. S. T. K. Chan and B. E. Larock, 'Fluid flows from axisymmetric orifices and valves', *Proc. ASCE, J. of the Hydraulics Division*, **99**, HY1, 81–97 (1973).
16. F. W. Steffen, J. G. McArdle and J. W. Coats, 'Performance characteristics of hemispherical target-type thrust reversers', *NACA RM E55E18*, June 3, 1955.
17. T. Sarpkaya, 'Of fluid mechanics and fluidics and of analysis and physical insight', *Proceedings of the 5th Cranfield Fluidics Conference*, **5**, 33–54 (1972).
18. F. W. Steffen, H. G. Krull and C. C. Clepluch, 'Preliminary investigation of several target-type thrust reversal devices', *NACA RM E53L15b*, March 12, 1954.

Author Index

The numbers in parentheses refer to entries in the Reference sections at the end of each chapter.

Kealy, C. D., 158(20), 164(20), 169(20)
Kelsey, S., 10(22)
Kellog, R. B., 15(32)
Kim, C. H., 220(8), 227(8)
King, I. P., 141(8), 158(14)
Kinsman, G., 119(1)
Koh, R. C. Y., 133(5)
Kosko, E., 61(13)
Krentos, V. D., 202(7)
Krull, H. G., 277(18)

Lamb, H., 133(1), 251(6)
Landweber, L., 219(3)
Larock, B. E., 266(15)
Lee, C. M., 220(10)
Lee, K. K., 124(14), 125(18)
Lee, K. N., 7(12), 10(18), 10(19)
Leendertse, J. J., 95(3), 98(3), 114(3), 115(3), 116(3)
Lennon, G. W., 95(6)
Leonard, J., 126(20)
Lew, H. S., 37(14)
Lewis, F. M., 219(2)
Lewis, R. W., 158(60), 158(61)
Li, C. Y., 158(41), 164(41)
Liao, K. H., 158(71)
Liggett, J. A., 120(4), 124(14)
Lindberg, G. M., 61(13), 68(16)
Liniger, W., 183(3), 188(3)
Longenbaugh, R. A., 158(51)
Loziuk, L., 127(22), 127(23)
Luthin, J. N., 165(81)
Lynn, P. P., 7(11)
Lytton, R. L., 167(99), 177(99)

Macagno, M., 219(3)
Maini, Y. N. T., 165(85)
Maraczek, G., 57(2)
Marcal, P. V., 74(11), 86(11)
Martin, H. C., 1(1), 9(1), 10(1), 12(1), 16(1), 57(1)
Matsumoto, K., 220(15)
Matsuura, Y., 220(14)
May, I., 86(19), 93(19)
Mayer, P., 158(3)
Mayer, P. G., 163(79)
McArdle, J. G., 274(16), 276(16), 277(16)
McCorquodale, J. A., 158(28), 160(28), 163(28), 164(28), 167(28), 171(28)
Mehaute, B. le, 133(2)
Melfi, D., 126(20)
Mills, K. G., 158(32), 164(32)

Milton, J. L., 166(94)
Missbach, A., 159(66)
Mitchell, A. R., 237(8)
Moltz, F. J., 157(2), 160(2), 166(2)
Moore, F. K., 119(2)
Morgenstern, N. R., 158(56), 158(57)
Morton, K. W., 166(90)
Morris, M. B., 104(11), 108(11)
Motz, L., 128(26)
Murray, D. W., 158(56), 158(57)

Nagaoaka, H., 158(54), 158(55)
Nair, K. D., 86(19), 93(19)
Nakagawa, K., 73(6), 77(6)
Nalluswami, M., 158(51)
Nath, B., 220(11)
Naylor, D. J., 158(5)
Neuman, S. P., 158(19), 158(25), 158(29), 162(19), 162(30), 162(77), 164(19), 164(29), 201(2), 201(10), 201(11), 202(9), 215(8)
Nemay-Nasser, S., 10(18), 10(19)
Newton, R. E., 220(16), 220(19), 226(19)
Ney, R. A., 20(37)
Noorishad, J., 158(45), 165(85)
Norrie, D. H., 266(14)
Norton, W. R., 141(8)

O'Brien, G. G., 166(91)
Oden, J. T., 6(5), 32(4), 57(5), 57(6), 57(9), 77(14), 158(62), 165(82), 165(84), 167(82), 177(84)
Ogilvie, T. F., 219(1)
Oliviera, E. R. A., 10(14)
Olson, M. D., 57(8), 60(8), 61(13), 68(16), 73(1)
Opsteegh, J. D., 234(6)
Orlob, G., 124(11), 141(18)
Owen, D. R. J., 7(12)

Palit, K., 46(20), 46(21)
Parekh, C. J., 50(25), 158(4), 158(29), 164(29), 201(4)
Patil, B. S., 256(8)
Paulling, J. R., 220(6), 227(6), 228(6), 229(6)
Pearson, J. R. A., 85(15)
Perzyna, P., 47(22)
Peters, J. C., 50(25), 158(29), 164(29), 201(4)
Pian, T. H. H., 10(20), 15(33)

Subject Index